系统能力培养（计算机网络）系列教材

计算机网络协议分析与实践

姚　烨　朱怡安　张黎翔　编著

电子工业出版社
Publishing House of Electronics Industry
北京·BEIJING

内 容 简 介

本书系统地介绍了计算机网络协议分析与实践的有关内容，主要包括：网络编程基础及环境配置，工业以太网、共享式以太网协议分析与实践，网络通信协议的设计与实践，以及网络层（ARP、IPv4/IPv6、ICMP）、传输层（UDP、TCP）、应用层（HTTP、FTP）等常用通信协议工作原理分析与实践等内容，基本涵盖了 TCP/IP 协议集的常用网络通信协议。本书实践环节基于主流开发环境和工具，不需要特殊的软、硬件平台投入，既方便学生课后实践，又方便教师组织实践教学活动。

本书内容系统性较强、结构清晰。在内容的组织上，本书强调知识的实用性，从网络通信协议三要素：语法、语义和同步关系三个角度分析 TCP/IP 协议集的常用网络通信协议工作原理和相关实现技术，对学生理解计算机网络通信协议基本理论，提高学生网络信息安全基本实践技能均有很大益处。

本书可作为高等院校相关专业学生的专业教材，也可作为相关技术人员的自学用书。

未经许可，不得以任何方式复制或抄袭本书之部分或全部内容。
版权所有，侵权必究。

图书在版编目（CIP）数据

计算机网络协议分析与实践 / 姚烨，朱怡安，张黎翔编著. — 北京：电子工业出版社，2021.1
ISBN 978-7-121-40250-0

Ⅰ.①计⋯ Ⅱ.①姚⋯ ②朱⋯ ③张⋯ Ⅲ.①计算机网络－通信协议－高等学校－教材 Ⅳ.①TN915.04

中国版本图书馆 CIP 数据核字（2020）第 256284 号

责任编辑：孟　宇　　　　特约编辑：田学清
印　　刷：大厂聚鑫印刷有限责任公司
装　　订：大厂聚鑫印刷有限责任公司
出版发行：电子工业出版社
　　　　　北京市海淀区万寿路 173 信箱　邮编：100036
开　　本：787×1092　1/16　印张：20　字数：500 千字
版　　次：2021 年 1 月第 1 版
印　　次：2021 年 7 月第 2 次印刷
定　　价：59.00 元

凡所购买电子工业出版社图书有缺损问题，请向购买书店调换。若书店售缺，请与本社发行部联系，联系及邮购电话：(010)88254888，88258888。
质量投诉请发邮件至 zlts@phei.com.cn，盗版侵权举报请发邮件至 dbqq@phei.com.cn。
本书咨询联系方式：mengyu@phei.com.cn。

前言

"计算机网络原理"作为计算机专业的核心课程,编者以问题为引导、以实践为抓手、以能力培养为核心,践行了讲网络不如做网络的原则,实现了知行统一的教育教学新模式。编者通过多年理论课程教学与实践发现,若理论课程教学以问题为牵引,以案例教学为核心,则更容易引发学生思考、培养学生分析问题的能力、启发学生思维、激发学生的学习兴趣。在实践方面,编者采用多层次实践教学方法:课程实验主要培养学生对计算机网络工作原理及通信协议和应用服务的验证与分析能力;课程项目主要以每节课的内容为核心,指导学生将每节课的核心内容用软件操作一遍,做到"所讲即所见";网络综合实验主要培养学生在网络工程实践、协议栈设计及实现等方面的能力。实践表明,以上方法可以较好地解决学生学习成绩断崖式下降的问题,提高学生网络实践的能力,提升学生对课程的认同感。

编者通过对计算机网络实践内容和经验的总结,形成了一套计算机网络课程实践教材(共 3 本)。本教材主要针对计算机网络协议分析与实践这一核心内容:第 1 章主要介绍网络编程基础及环境配置;第 2 章主要介绍工业以太网协议分析与实践;第 3 章主要介绍共享式以太网协议分析与实践;第 4 章主要介绍网络通信协议的设计与实践;第 5 章、第 6 章、第 7 章分别介绍网络层 ARP、IP 和 ICMP 协议分析与实践;第 8 章、第 9 章分别介绍传输层 UDP 协议和 TCP 协议分析与实践;第 10 章介绍应用层协议分析与实践;第 11 章介绍 IPv6 协议分析与实践。

本书在编写过程中,得到了西北工业大学计算机学院的老师和同学们的支持和指导,在此表示感谢!

由于编者水平有限,书中难免有不妥之处,恳请读者批评指正。

编 者
2020 年 11 月

目 录

第1章 网络编程基础及环境配置 ... 1
1.1 引言 ... 1
1.2 Raw Socket 网络编程基础 ... 2
1.2.1 Raw Socket 的创建和初始化 ... 3
1.2.2 Raw Socket 网络编程主要函数 ... 4
1.2.3 简单校验和计算方法 ... 6
1.2.4 创建不同层次协议数据单元首部数据结构 ... 7
1.3 Winpcap 网络编程基础 ... 10
1.3.1 数据包捕获的基本过程 ... 11
1.3.2 Winpcap 驱动内部工作原理 ... 12
1.3.3 利用 winpcap 对数据包进行捕获和过滤 ... 16
1.3.4 主要数据结构和接口函数 ... 17
1.4 Winpcap 环境搭建 ... 22
1.4.1 Visual Studio 10.0 实验环境搭建 ... 22
1.4.2 VC++6.0 环境配置 ... 24

第2章 工业以太网协议分析与实践 ... 25
2.1 引言 ... 25
2.2 工业以太网协议工作原理 ... 25
2.2.1 工业以太网协议语法 ... 26
2.2.2 工业以太网协议语义 ... 28
2.2.3 工业以太网协议时序关系 ... 28
2.3 工业以太网发送数据帧 ... 28
2.4 工业以太网接收数据帧 ... 31

第3章 共享式以太网协议分析与实践 ... 37
3.1 引言 ... 37
3.2 共享式以太网的工作原理 ... 40
3.2.1 共享式以太网数据帧语法及语义 ... 41
3.2.2 共享式以太网数据帧的发送及其过程分析 ... 42
3.3 共享式以太网数据链路层协议工作效率分析 ... 44
3.4 网络适配器 MAC 地址 ... 45
3.5 工业以太网数据帧发送和接收流程 ... 46

3.5.1 工业以太网数据帧发送流程 46
3.5.2 工业以太网数据帧接收流程 46
3.5.3 MAC 子层与相邻层的接口 47
3.6 共享式以太网数据帧发送源程序 48

第 4 章 网络通信协议的设计与实践 52
4.1 引言 52
4.2 网络通信协议可靠性原理 52
4.2.1 检错与纠错机制 52
4.2.2 流量控制机制 55
4.3 数据链路层通信协议设计 60
4.3.1 数据链路层通信协议设计要求 60
4.3.2 数据链路层通信协议语法设计 60
4.3.3 数据链路层通信协议语义设计 60
4.3.4 数据链路层通信协议同步机制设计 61
4.4 数据链路层可靠通信协议实现 63
4.4.1 编程接口 Winpcap 63
4.4.2 网络通信协议并发机制实现技术 64
4.4.3 差错控制机制实现技术 66
4.4.4 发送方线程与接收方线程实现技术 66
4.4.5 停止-等待协议实现技术 79
4.4.6 后退 N 帧协议实现技术 108
4.4.7 选择重传协议实现技术 138

第 5 章 网络层 ARP 协议分析与实践 172
5.1 概述 172
5.2 ARP 协议工作原理 173
5.2.1 ARP 协议语法 173
5.2.2 ARP 协议语义 174
5.2.3 ARP 协议时序关系 175
5.3 ARP 协议发送报文 177
5.4 ARP 协议接收报文 179

第 6 章 网络层 IP 协议分析与实践 184
6.1 引言 184
6.2 IP 协议工作原理 186
6.2.1 IP 协议语法 186
6.2.2 IP 协议语义 188
6.2.3 IP 协议时序关系 188

6.3　IP 协议发送 IP 分组 193
6.4　IP 协议接收 IP 分组 195

第 7 章　网络层 ICMP 协议分析与实践 200
7.1　引言 200
7.2　ICMP 协议工作原理 200
　　7.2.1　ICMP 协议语法 200
　　7.2.2　ICMP 差错报告报文语义及同步关系 201
　　7.2.3　ICMP 控制报文语义及同步关系 203
　　7.2.4　ICMP 查询报文语义及同步关系 203
7.3　ping 命令实现分析 204
7.4　Tracert 命令设计与实现 213
7.5　ICMP 协议发送 ICMP ECHO 请求报文 224
7.6　ICMP 协议接收 ICMP ECHO 请求报文 227

第 8 章　传输层 UDP 协议分析与实践 231
8.1　引言 231
8.2　UDP 协议工作原理 231
　　8.2.1　UDP 协议语法及语义 231
　　8.2.2　UDP 协议时序关系 232
8.3　UDP 协议发送 UDP 用户数据报 233
8.4　UDP 协议接收 UDP 用户数据报 236

第 9 章　传输层 TCP 协议分析与实践 238
9.1　TCP 协议概述 238
9.2　TCP 协议工作原理 242
　　9.2.1　TCP 协议语法及语义 242
　　9.2.2　TCP 协议通信的时序关系 246
9.3　TCP 协议发送数据段 263
9.4　TCP 协议接收数据段 266

第 10 章　应用层协议分析与实践 271
10.1　引言 271
10.2　HTTP 协议工作原理 272
　　10.2.1　统一资源定位符 274
　　10.2.2　HTTP 1.0 协议的主要特点 274
　　10.2.3　Web 代理服务器 276
　　10.2.4　HTTP 报文的语法和语义 277
　　10.2.5　Cookie 工作原理 280
10.3　万维网文档 282

10.3.1 超文本标记语言 ··· 282
10.3.2 动态文档 ··· 284
10.3.3 活动文档 ··· 284
10.4 HTTP 协议客户端实现 ··· 285
10.5 FTP 协议工作原理 ··· 287
10.5.1 FTP 协议概述 ··· 287
10.5.2 FTP 协议工作模式 ··· 288
10.5.3 FTP 协议命令 ··· 289
10.6 FTP 协议客户端实现 ··· 291

第 11 章 IPv6 协议分析与实践 ··· 297
11.1 引言 ··· 297
11.2 IPv6 协议工作原理 ··· 297
11.2.1 IPv6 协议语法及语义 ··· 297
11.2.2 IPv6 协议的地址空间 ··· 300
11.3 IPv6 协议地址空间的分配 ··· 301
11.4 从 IPv4 协议向 IPv6 协议过渡机制 ··· 303
11.5 IPv6 协议发送分组 ··· 304
11.6 IPv6 协议接收分组 ··· 308

参考文献 ··· 311

第 1 章 网络编程基础及环境配置

1.1 引　　言

20 世纪 80 年代，美国政府的高级研究工程机构为加利福尼亚大学伯克利分校提供资金，委托其在 UNIX 系统下开发 TCP/IP 编程接口，其工作成果就是 Socket，一般称为套接字。Socket 屏蔽了底层网络（数据链路层和物理层）和操作系统的差异，使得任意两台安装 TCP/IP 协议集的计算机之间能够通过该接口实现通信。

20 世纪 90 年代，微软公司联合其他公司，如 IBM 公司等，制定了 Windows 环境下的网络编程接口：Windows Socket（简称 Winsock）规范。目前，Windows 环境下的编程都是基于 Winsock 规范的，包括 16 位和 32 位两种网络编程接口。Winsock 的开发工具在 Borland C++ 和 Visual C++等基于 C 语言的编译器中均有提供，主要由 Winsock.h 和 winsock.dll 等提供接口函数。Winsock 的发展经历了两个版本：Winsock1.1 和 Winsock2.0，后者是前者的扩展，并向后兼容。在 Visual C++环境下，主要使用 CAsyncSocket 和 CSocket 类，其中 CAsyncSocket 类利用面向对象技术对 Winsock 接口进行封装，CSocket 类是 CAsyncSocket 的子类，二者都是 MFC 类库中的类。

Winsock 在高层支持流通信和数据报通信两种模式，前者用于 TCP 通信编程，后者主要用于 UDP 应用编程。由于网络负载随时会发生变化，特别是网络有时会发生拥塞现象，使得网络通信的收发函数无法及时返回，这种现象称为阻塞。为了对可能发生阻塞现象的函数进行处理，Winsock 提供了两种处理方式：阻塞方式和非阻塞方式，前者又称为同步方式，后者又称为异步方式。在同步方式下，收发函数被调用后要等数据传输完毕或出错时才能返回，在此期间要等待操作系统操作完成，才能做任何其他事情。对于异步方式，收发函数被调用后立即返回，当操作系统数据传输完毕，由 Winsock 接口负责给应用程序发送一个消息，通知数据传输完成。应用层程序可根据传输的消息参数判断本次通信是否成功。一般编程时建议采用异步方式，以提高系统的工作效率。例如，在 MFC 编程环境中，SendMessage()是同步函数，当其发送消息到消息队列后，需要等待消息执行完成后才能返回；而 PostMessage()是异步函数，不管发送的消息是否被处理及处理结果如何，发送完消息后立即返回。Winsock 提供了 5 种 I/O 模型来处理异步通信模式，分别为 Select（选择）、WSAAsyncSelect（异步选择）、WSAEventSelect（事件选择）、overlapped（重叠）、completion port（完成端口）。

Winsock1.1 接口称为网络编程规范，其提供了超强和十分灵活的 API，但存在的问题是，它不像 Berkerly socket 模型可以支持多个协议，而是仅支持 TCP/IP 协议集；Winsock2.0 对 Winsock1.1 进行了扩展，可透明地支持多种协议，如 TCP/IP、ATM、IPX/SPX 和 DECNet 等，并兼容 Winsock1.1 应用程序。Winsock2.0 提供的新功能如下。

（1）多重协议支持：支持多种网络通信协议。
（2）多重命名空间：支持服务和解析域名选择协议。
（3）分散和聚集：可从多个缓存空间接收和发送数据。
（4）重叠的 I/O 和事件对象：支持异步通信模式。
（5）服务质量：根据 Socket 参数协商跟踪网络带宽。
（6）多点传输和条件接收：可选择性确定是否接收。
（7）Socket 共享：多个进程共享一个 Socket 句柄。
（8）扩展性：用户可自主增加新的 API，扩展新的服务。

WinInet 编程接口也称为 Win32 Internet Extensions，是由 Winsock 封装和扩展的，主要用于应用层客户端编程，如 Gopher、HTTP、FTP 等应用层协议。虽然用户可以利用 Winsock 接口实现更高级的应用层协议，但 WinInet 提供了一组函数，采用了 Win32 API，使实现应用层客户端编程很简单。Visual C++提供了两种使用 WinInet 编程接口的方式：一种是一组 Win32 API；另一种是 MFC 封装的 WinInet 类。MFC 提供了 3 个由 CStdioFile 派生的子类，分别为 CInternet、CHttpFile、CGopherFile。

1.2 Raw Socket 网络编程基础

用户为了灵活构造自己设计的协议数据单元，并通过网络接口发送，一般会用到 Raw Socket（原始套接字），目的是实现对网络数据包进行一定程度的控制。Raw Socket 常用于开发简单的网络监测、网络报文捕获与分析、网络攻击、入侵检测、流量监控等应用和工具。一个基于 Raw Socket 网络程序具有 Windows Sockets DLL 初始化和释放功能、将用户输入地址转化为 struct sockaddr_in 数据结构的地址转化功能、创建 Raw Socket 和释放 Raw Socket 功能、对 Raw Socket 的发送和接收选项进行配置的功能、构建数据包首部并发送功能、接收数据包并解析功能、发生错误时释放 Raw Socket 和资源回收功能等。

基于 Raw Socket 的数据包发送过程如下。
（1）Windows Sockets DLL 初始化，指定使用的版本号。
（2）创建 Raw Socket，选择使用 Raw Socket，根据需要设置 IP 控制选项。
（3）确定目的 IP 地址和端口号。
（4）构造发送的 IP 分组（用户可灵活定制 IP 分组以上的协议数据单元）。
（5）发送 IP 分组。
（6）关闭 Raw Socket。
（7）释放 Windows Sockets DLL 资源。

基于 Raw Socket 的数据包的接收过程如下。
（1）Windows Sockets DLL 初始化，指定使用的版本号。

（2）创建 Raw Socket，选择使用 Raw Socket，根据需要设置特定的协议类型。
（3）根据需要设定接收选项。
（4）接收数据包并进行过滤。
（6）关闭 Raw Socket。
（7）释放 Windows Sockets DLL 资源。

1.2.1　Raw Socket 的创建和初始化

输入参数：
BOOL bSendflag：发送控制选项。
BOOL bRecvflag：接收控制选项。
int iProtocolType：协议类型。
socketaddr_in *ptrLocalIP：本地 IP 地址的数据结构，作为返回值。
输出：
返回创建并初始化好的 Raw Socket 句柄。

```
SOCKET Creat_rawsocket( BOOL bSendflag, BOOL bRecvflag, int iProtocol,
socketaddr_in *ptrLocalIP )
{
    SOCKET Raw_Socket;
    Int iResult =0, in =0, i=0;
    Struct hostent *localhost;
    Char hostname[DEFAULT_NAME_LENGTH];
    Struct in_addr addr;
    DWORD dwBufferLen[10],optional =1, dwReturned =0;
    WORD wDllVersion;
    WSADATA wsaData;
    //装载并初始化 Windows Sockets DLL，版本号为 2.2
    wDllVersion = MAKEWORD(2,2);
    iResult = WSAStartup(wDllVersion, &wsaData);
    if (iResult != 0)
    {
        Printf("WSAStartup function error : %ld\n", WSAGetLastError());
        Return -1;
    }
    //创建 Raw Socket
    Raw_Socket = socket(AF_INET, SOCK_RAW, iProtocol);
    If (Raw_Socket == INVALID_SOCKET)
    {
        Printf("raw socket create error ,error number : %ld\n", WSGetLastError()));
        WSACleanup();
        Return -1
    }

    If (bSendflag == TRUE)
```

```
    {
        //设置 IP_HDRINCL 表示需要构造 IP 分组首部，需要#include "ws2tcpip.h"
        Setsocktopt(Raw_socket, IPPROTO_IP,IP_HDRINCL,(char *) &bSendflag,sizeof
                    (bSendflag));
    }

    If (bRecvflag == TRUE)
    {
         Memset(HostName, 0, DEFAULT_NAMELENGTH);
        //获取主机名称
        Gethostname (HostName , sizeof (HostName));
        //获取主机 IP 地址
        bLocalIP = gethostbyname(HostName);
        while(localIP ->h_addr_list[i] != 0)
        {
            Addr.s_addr = *(u_long *) local->h_addr_list[i++];
            Printf("please select the network interface number : ");
            Scanf_s("%d",&in);
            Memset(ptrLocalIP, 0, sizeof(sockaddr_in));
            Memcpy(&ptrLcalIP->sin_addr.S_un.S_addr,
                local->h_addr_list[in-1],sizeof(ptrLocalIP->sin_addr.S_un.S_addr)));
            ptrLocalIP->sin_family = AF_INET;
            ptrLocalIP ->sin_port = 0;
            bind(Raw_socket,(struct sockaddr *) ptrLocalIP, sizeof(sockaddr_in));
            //设置 SIO_RCVALL 表示接收工作接口上所有报文
            WSAIoctl(Raw_Socket, SIO_RCVALL, &optional, sizeof(optional),
                ?&dwBufferLen, sizeof(dwBufferLen), &dwReturned, NULL, NULL);
        }
    }
    Return Raw_Socket;
}
```

1.2.2 Raw Socket 网络编程主要函数

1）创建 Raw Socket

可以用 socket()或 WSASocket()函数来创建 Raw Socket，因为 Raw Socket 能直接控制底层协议，所以只有属于"管理员"组的成员，才有权创建 Raw Socket。下面是用 socket()函数创建 Raw Socket 的代码。

```
SOCKET Socketid;
Socketid = socket (AF_INET, SOCK_RAW, IPPROTO_UDP);
```

上述创建 Raw Socket 的代码使用的是 UDP 协议，如果要使用其他的协议，如 ICMP、IGMP、IP 等协议，则只需要把相应的参数改为 IPPROTO_ICM、IPPROTO_IGMP、IPPROTO_IP 就可以了。另外，IPPROTO_UDP、IPPROTO_IP、IPPROTO_RAW 这几个协议标志要求使用套接字选项 IP_HDRINCL，而目前只有 Windows 2000 和 Windows XP 提供

了对 IP_HDRINCL 的支持，这意味着在 Windows 2000 以下平台创建 Raw Socket 时，是不能使用 IP、UDP、TCP 等协议的。

像其他类型的 Socket 一样，Raw Socket 的创建非常简单，直接使用 socket()函数进行创建就可以了。

SOCKET_STREAM：流式套接字。

SOCKET_DGRAM：用户数据报（UDP）。

SOCKET_RAW：原始套接字。

IPPROTO_IP：IP 协议。

IPPROTO_ICMP：INTERNET 控制消息协议，配合 Raw Socket 可以实现 ping 的功能。

IPPROTO_IGMP：INTERNET 网关服务协议，在多播中用到。

在 AF_INET 地址族下，有 SOCK_STREAM、SOCK_DGRAM、SOCK_RAW 3 种套接字类型。SOCK_STREAM 也就是通常所说的 TCP 协议，而 SOCK_DGRAM 则是通常所说的 UDP 协议，而 SOCK_RAW 则是用于提供一些较低级的控制的。socket()函数中第 3 个参数依赖于第 2 个参数，用于指定套接字用的特定协议，设为 0 表示使用默认的协议。

```
int socketfd = socket(AF_INET,SOCK_RAW,IPPROTO_ICMP); /*在网络层使用的Raw Socket
/*在链路层使用的Raw Socket*/
int socketfd = socket(PF_PACKET,SOCK_RAW,htons(ETH_P_IP));
```

在指定协议的时候，不能向其他套接字一样简单地指定为 0（IPPROTO_IP），因为其他套接字会根据套接字类型自动选择协议。例如，stream 类型的协议会选择 TCP 协议，而 Raw Socket 不行。这些协议的宏定义在 UNIX 系统下定义在"netinet/in.h"文件里，当然要使用这些协议还需要内核层对该协议的支持。

```
Raw_Socket = socket(AF_INET, SOCK_RAW, iProtocol)
```

2）设置首部选项

创建 Raw Socket 后，就要设置套接字选项了，这要通过 setsocketopt()函数来实现。setsocketopt()函数的声明如下。

```
int setsocketopt (
SOCKET s,
int level,
int optname,
const char FAR *optval,
int optlen
);
```

在上述声明中，参数 s 是标识套接字的描述字，要注意的是，选项对这个套接字必须是有效的。参数 level 表明选项定义的层次，对 TCP/IP 协议集而言，其支持 SOL_SOCKET、IPPROTO_IP 和 IPPROTO_CP 层次；参数 optname 是需要设置的选项名，这些选项名是在 Winsock 头文件内定义的常数值；参数 optval 是一个指针，它指向存放选项名的缓存区；参数 optlen 指示 optval 缓存区的长度。

```
Setsocktopt(Raw_socket, IPPROTO_IP,IP_HDRINCL,(char *) &bSendflag,sizeof
(bSendflag))
```

使用 setsockopt()函数来设置套接字选项，其中 IP_HDRINCL 用来设置是否手动处理 IP 包头，如果设置为真，那么需要自己创建 IP 包头，然后发送；如果没有设置，那么系统会自动为 Raw Socket 设置 IP 包头，并附加在自己的数据之前。当然，使用 Raw Socket 接收的数据包总是包含 IP 包头的。正是因为系统提供了这样的操作，使得用户在 root 权限下可以使用虚假源 IP 地址发送数据。

3）设置控制选项

```
int WSAAPI WSAIoctl(SOCKET s,            //一个套接字的句柄
DWORD dwIoControlCode,                   //将进行操作的控制代码
LPVOID lpvInBuffer,                      //输入缓存区的地址
DWORD cbInBuffer ,                       //输出缓存区的地址
LPVOID lpvOutBuffer,                     //输入缓存区的大小
DWORD cbOutBuffer,                       //输出缓存区的大小
LPDWORD lpcbBytesReturned,               //输出实际字节数的地址
LPWSAOVERLAPPED lpOverlapped,            //WSAOVERLAPPED 结构的地址
LPWSAOVERLAPPED lpCompletionRoutine      //一个指向操作结束后调用的例程指针
)
```

返回值：

调用成功后，WSAIoctl()函数返回 0，否则，将返回 SOCKET_ERROR（错误），应用程序可通过 WSAGetLastError()来获取相应的错误代码。例如：

```
WSAIoctl(Raw_Socket, SIO_RCVALL, &optional, sizeof(optional), ?&dwBufferLen, sizeof(dwBufferLen), &dwReturned, NULL, NULL)
```

1.2.3 简单校验和计算方法

TCP、UDP、IP 等协议在计算校验和时，采用简单校验和计算方法。TCP 协议和 UDP 协议的计算校验和方法：将伪首部、首部和数据部分内容按照 16 比特位对齐，采用二进制加法相加，得到的和取反即校验结果；IP 协议采用同样的计算方法，但仅对 IP 分组首部（包括基本首部和可变首部）内容计算校验和。

```
USHORT check_sum(USHORT *ptrBuffer, int iSize)
{
    Unsigned long ulResultCheck = 0;
    While (iSize > 1)
    {
        ulResultCheck += *ptrBuffer ++;
        iSize   -= sizeof(USHORT);
    }
    If(iSize)
    {
        ulResultCheck += *(UCHAR*) ptrBuffer;
    }
    ulResultCheck = (ulResultCheck >> 16) + (ulResultCheck & 0xffff);
    ulResultCheck += (ulResultCheck >> 16);
    return (USHORT) (~ulResultCheck);
}
```

1.2.4　创建不同层次协议数据单元首部数据结构

1）TCP 首部数据结构

TCP 首部由基本首部（20 字节）和选项部分（0~40 字节）构成，其中基本首部主要包括：源端口号、目的端口号、序号、确认序号、4 比特位首部长度（以 4 个字节为计算单位）、6 比特位保留字段、6 比特位标志字段、窗口大小字段、校验和和紧急指针字段等，如图 1-1 所示。

图 1-1　TCP 首部数据结构示意图

TCP 首部数据结构的定义如下。

```
Typedef struct   tagTCPHeader
{
    USHORT       source_port;
    USHORT       dest_port;
    ULONG        sequence;
    ULong        ack_number;
    BYTE         header_len;
    BYTE         flags;
    USHORT       recv_win_size;
    USHORT       check_sum;
    USHORT       urgent_size;
}TCPHeader, *ptr TCPHeader;
//TCP 标志位定义
#define   TCP_FIN      0X01      //释放 TCP 连接请求
#define   TCP_SYN      0X02      //TCP 连接请求
#define   TCP_REST     0X04      //TCP 连接复位
#define   TCP_PUSH     0X08      //PUSH 操作
#define   TCP_ACK      0X10      //ACK 应答有效
#define   TCP_URGENT   0X20      //紧急指针有效
```

2）UDP 用户数据报首部数据结构

UDP 用户数据报由首部和数据两部分构成，其中首部共 8 个字节，主要包括源端口号、目的端口号、长度和校验和字段；UDP 用户数据报在计算简单校验和时，会临时产生一个伪首部，所以简单校验和计算范围主要包括：伪首部+首部+数据；伪首部共 12 个字节，主要包括源 IP 地址、目的 IP 地址、1 个字节保留字段（填充域 0）、UDP 协议编号字段（值为 17，占用 1 个字节）、UDP 用户数据报长度（包括首部和数据两部分长度）。UDP 用户数据报首部及伪首部结构如图 1-2 所示。

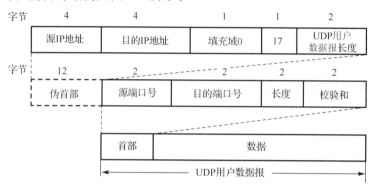

图 1-2　UDP 用户数据报首部及伪首部结构

UDP 用户数据报首部数据结构的定义如下。

```
//定义 UDP 用户数据报伪首部，在计算简单校验和时需要
Type struct tagUDPPseudoHeader
{
    ULONG   IPSource_Address;
    ULONG   IPDest_Address;
    UCHAR   zero;
    UCHAR   ProtocolType;
    USHORT  UDP_Length;
} UDPPseHeader, * ptrUDPPseHeader;
//定义 UDP 用户数据报首部
Type struct tagUDPHeader
{
    USHORT  UDP_source_port;
    USHORT  USP_dest_port;
    USHORT  UDP_length;
    USHORT  UDP_checksum;
} UDPHeader, *ptrUDPHeader;
```

3）ICMP 数据单元首部数据结构

ICMP 协议作为网络层协议之一，其主要功能是辅助 IP 协议完成差错报告、拥塞控制、重路由选择及网络信息查询。该协议产生的协议数据单元由首部和数据两部分构成，首部主要包括 1 个字节类型字段、1 个字节代码字段（或子类型字段）、2 个字节的校验和字段及 4 个字节保留字段（未用），具体内容如图 1-3 所示。

图 1-3　ICMP 数据单元首部数据结构示意图

ICMP 数据单元首部数据结构的定义如下。

```
Typedef    struct tagICMPHeader
{
    UCHAR   type;
    UCHAR   code;
    UCHAR   checksum;
    UCHAR   id;
    UCHAR   sequence;
}ICMPHeader, *ptrICMPHeader
```

4）IP 分组首部数据结构

IP 协议产生的协议单位称为 IP 分组或 IP 包，IP 分组由首部和数据两部分构成。IP 分组的首部又由基本首部（20 字节）和可变首部（0~40 字节）构成。IP 分组的基本首部主要包括版本号、首部长度、区分服务、总长度、标识、标志（3 比特位）、片偏移、生存时间、协议类型、首部校验和、源 IP 地址和目的 IP 地址；IP 分组的可变首部主要包括可选字段和填充字段。IP 分组首部数据结构示意图如图 1-4 所示。

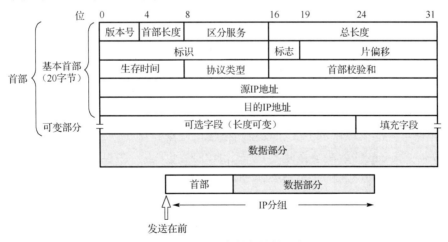

图 1-4　IP 分组首部数据结构示意图

IP 分组首部数据结构的定义如下。

```
Typedef    struct tagIPHeader
{
```

```
    UCHAR   header_len: 4;
    UCHAR   version: 4;
    UCHAR   tos;
    USHORT  total_len;
    USHORT  id;
    USHORT  flag_offset;
    UCHAR   ttl;
    UCHAR   protocol;
    USHORT  check_sum;
    ULONG   source_ipaddress;
    ULONG   dest_ipaddress;
}IPHeader, *ptrIPHeader
```

1.3 Winpcap 网络编程基础

Winpcap 是 Windows packet capture 的缩写,提供 Windows 平台下可访问数据链路层接口的开源库,可实现在数据链路层的数据帧的构造、捕获和分析;Winpcap 接口是由加利福尼亚大学和 lawrence berkeley 实验室联合开发的一个免费、开源项目,目前最新版本是 Winpcap4.1.2,支持 X86 和 X64 两个环境,其官方网站为 http://www.winpcap.org,用户可以在其主页上下载 Windows 驱动、源代码和开发文档。Windows 在内核层实现了数据包的捕获和过滤,可以独立于 TCP/IP 协议栈实施原始数据包的构造、发送和接收。基于 Winpcap 接口可实现以下功能。

(1) 直接在网络接口上捕获或发送数据包。

(2) 在将捕获的数据包交付给应用程序前,根据用户指定的规则实现核心层数据包过滤。

(3) 收集并统计数据流量。

(4) 实现软交换和软路由。

(5) 用于入侵检测系统。

基于 Winpcap 网络编程体系结构包含一个内核层数据包过滤器(NPF Device Driver)、底层动态链接库(Packet.dll)、高层并独立于系统的库(Wpcap.dll),如图 1-5 所示。为了访问网络上传输的原始数据,一个捕获系统需要绕过操作系统的协议栈,一部分程序采用运行于操作系统的内核中的方式,来与网络接口驱动直接交互,而该部分程序与操作系统密切相关。Winpcap 的解决方案是实现采用 NPF 驱动程序,为 Windows 95、Windows 98、Windows ME、Windows NT 4、Windows 2000 与 Windows XP 等不同操作系统提供不同版本的驱动程序。首先,这些驱动程序提供了数据包捕获与发送的基本特性,也提供了如可编程的过滤系统与监控引擎等更高级的功能特性,前者可用于限制一个捕获会话,只捕获特定的网络数据包(如只捕获一个特定主机生成的 FTP 数据包),后者提供了一个简单但功能强的方式来获取网络流量的统计信息,如获取网络负载或两个主机间交换数据的数量等信息。其次,捕获系统必须提供一个接口,使用户层应用程序可使用内核驱动提供的这些特性。Winpcap 提供两个不同的库:Packet.dll 与 Wpcap.dll。第一个库提供了一个底层的 API,可用来直接访问驱动程序的函数,库中提供的接口函数的调用与操作系统类型无关;

第二个库提供了功能更强、更高层的捕获函数接口，并与 UNIX 捕获库 libpcap 兼容。这些函数接口使数据包捕获能够独立于底层操作系统及网络接口硬件。

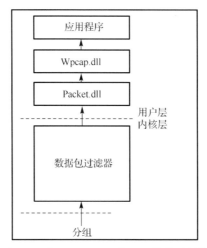

图 1-5　基于 Winpcap 编程网络体系结构

1.3.1　数据包捕获的基本过程

基于 Winpcap 从网络上捕获一个数据包，然后传递给应用程序，其调用的组件和通信流程如图 1-6 和图 1-7 所示。

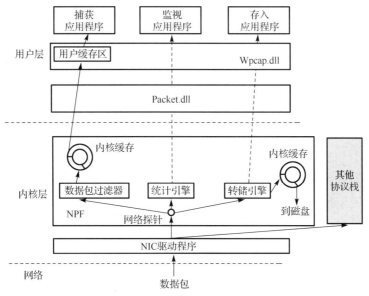

图 1-6　NPF 内部结构示意图

1）NIC 与 NIC 设备驱动

一般 NIC（Network Interface Card，NIC）（网卡）的内存大小通常限制为几千字节，这些内存在全连接速度（full link speed）下需要满足数据包的接收与发送条件。NIC 在数据包被存储在内存中时，需要执行一些初步的检查，如果出现 CRC 错误、碎片帧或无效帧等，则可以立即丢弃。

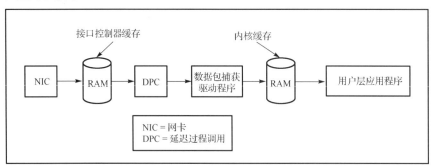

图 1-7　网络数据通信过程示意图

一个有效数据帧被 NIC 接收后，将对总线控制器产生一个总线数据传输请求。此时，NIC 控制总线传递数据帧给卷积主内存中的 NIC 缓存区中（见图 1-7），然后释放总线，产生一个硬件中断给高级可编程中断控制器（Advanced Programmable Interrupt Controller，APIC）芯片，该芯片唤醒操作系统的中断处理例程（OS interrupt handling routine），中断处理例程负责触发 NIC 驱动程序的中断服务程序（ISR）并完成数据传递任务。

NIC 驱动程序的 ISR 最基本的功能是检查该中断是否是它自己要处理的，因为在 X86 机器中，一个中断可被多个设备共享，并根据需要做出正确的应答，然后 ISR 调度一个较低优先级的函数，称作延迟过程调用（DPC），该函数稍后需要处理硬件请求与通告上层驱动程序（如协议层的驱动程序、数据包捕获驱动程序）一个数据包被接收了。当没有中断被挂起时，CPU 将处理 DPC 例程。当 NIC 驱动程序正在执行处理时，来自 NIC 的中断会被禁用，因为在处理下一个服务前，上一个数据包的处理必须完成。由于调用中断服务是一个耗费资源很大的操作，一般 NIC 允许多个数据包被送入一个中断的上下文中，因此每次激活上层驱动程序都能够处理多个数据包。

2）数据包捕获驱动

数据包捕获组件通常对其他的软件模块，如协议栈等是透明的，并不对标准的系统行为带来影响。当 NIC 接收到一个数据包时，NIC 驱动程序会将该消息通告给回调函数 Tap()，由回调函数 Tap()对该数据包按照过滤规则确定是丢弃还是接收。Tap()函数需要完成的基本功能是对数据包进行过滤。在 Win32 平台下，数据包捕获组件通常作为一个网络协议驱动程序被实现。

3）用户层的接口支持

网络必须给上层应用提供一个接口，使用户层应用程序可使用数据包捕获组件提供的功能；一般通过给用户层提供易于使用的库来实现。Winpcap 提供 Packet.dll 与 Wpcap.dll 两个库，这两个库使应用程序对数据包捕获组件提供的功能的使用能独立于底层操作系统及网络接口硬件。

1.3.2　Winpcap 驱动内部工作原理

NPF 是 Winpcap 的一个基本核心组件，用来处理网络上传输的数据包，并对用户层传输的数据包进行实时捕获、发送与分析。

1. 网络驱动程序接口规范

网络驱动程序接口规范（NDIS）定义了网络适配器，其主要功能是管理网络适配器的驱动程序与协议驱动程序之间的通信。NDIS 的主要目的是承担一个封装层，允许协议驱动程序不依赖特定网络适配器或特定 Win32 操作系统发送和接收网络（局域网或广域网）上的数据包。NDIS 支持 3 种类型的网络驱动程序。

1）NIC 驱动程序

NIC 驱动程序直接管理网络接口（网卡）。NIC 驱动程序在其下层直接与硬件交互，在其上层提供一个接口，允许高层或驱动程序在网络上发送数据包、处理中断、复位网卡、停止 NIC 驱动程序、请求与设置驱动程序的操作特性。NIC 驱动程序可以是微端口（miniport）驱动程序或全 NIC 驱动程序。微端口驱动程序仅针对特定的硬件操作来管理网络接口，包括发送与接收网络接口上的数据，而针对所有底层 NIC 驱动程序通用的操作，如同步管理等，则交由 NICS 提供服务；微端口驱动程序不能直接调用操作系统的例行程序，其与操作系统的接口是 NDIS。微端口驱动程序仅把数据包向上传递给 NDIS，并由 NDIS 将这些数据包传递给合适的协议。全 NIC 驱动程序既可以对特定硬件执行操作，也可以执行所有同步与排队等通常由 NDIS 来完成的操作；全 NIC 驱动程序要维护自己的绑定信息以标识接收的数据来源。

2）中间层驱动程序

中间层驱动程序是位于上层驱动程序（如协议驱动程序）与微端口驱动程序之间的接口。对上层驱动程序而言，中间层驱动程序与微端口驱动程序相似。对微端口驱动程序而言，中间层驱动程序与协议驱动程序相似。一个中间层驱动程序可以位于另一个中间层驱动程序的上层，但是这样的分层对系统性能有负面影响。开发中间层驱动程序的主要原因是，要在一个已有的遗留中间层驱动程序与一个微端口驱动程序之间实现介质转换任务，微端口驱动程序通常将 NIC 作为一种中间层驱动程序，并采用未知的新介质类型来管理。因此，一个中间层驱动程序能够从局域网协议转换到 ATM 协议；一个中间层驱动程序不能与用户层的应用程序通信，只能与其他 NDIS 驱动程序通信。

3）传输层驱动程序

传输层驱动程序是一种协议驱动程序，其目的是实现网络协议栈功能，如 IPX/SPX 或 TCP/IP 协议栈等，其一般工作在一个或多个网卡上层。一个协议驱动程序为其上层的应用层程序提供服务，同时与一个或多个 NIC 驱动程序或其下层的中间层 NDIS 驱动程序建立关联关系。NPF 作为一个协议驱动程序，从性能的角度来讲可能并不是最好的选择，但是其允许与 MAC 子层之间相互独立，同时能完全访问原始流量。NPF 在 NDIS 中的位置如图 1-8 所示。

在正常情况下，NPF 与操作系统采用异步交互方式，这意味着传输层驱动程序提供了一个回调函数集，在一些操作需要由 NPF 处理时，它们被系统调用。NPF 为上层应用程序提供了 I/O 操作接口，如 open、close、read、write、ioctl 等函数。NPF 与 NDIS 也采用异步交互方式，如一个新数据包到来的事件可通过一个回调函数，如 Packet_tap() 来通知 NPF。此外，NDIS 与 NIC 驱动程序的交互总是依靠非阻塞函数发生的。当 NPF 调用一个 NDIS 函数时，该调用结束后会立即返回；当调用结束后，NDIS 调用一个特定的 NPF 回调函数来通知该函数已经完成。驱动程序为底层提供的任何操作，如发送数据包、对 NIC 设置或请求参数等，均会提供回调函数。

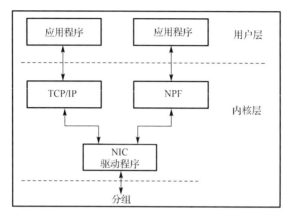

图 1-8　NDIS 中的 NPF 结构示意图

2．NPF 的主要功能

NPF 能够执行许多操作，如数据包捕获、数据包发送、网络监视、转储到磁盘等。

1）数据包捕获

NPF 最重要的操作是数据包捕获。在数据包捕获过程中，驱动程序利用网络接口嗅探数据包，并将数据包完好无缺地传递给用户层应用程序。数据包的捕获过程依赖两个主要组件：数据包过滤器和循环缓存区。

数据包过滤器：数据包过滤器决定一个到来的数据包是否需要被接收并传递给一个应用程序。大多数使用数据包过滤器的应用程序会拒收比它所能接收的多得多的数据包，因此，为了提升应用系统性能，一般采用 BSD 提供的数据包过滤器。如果一个虚拟处理器能够执行过滤程序，则该程序一般采用伪汇编程序实现并在用户层创建。应用程序一般采用用户定义的数据包过滤器。

循环缓存区：用来缓存数据包并避免丢包。存储在循环缓存区的数据包带有一个报文首部，首部字段包括地址、时间戳、数据包大小等信息。在单个读操作中，一组数据包能从 NPF 复制到应用程序，目的是提高性能、减少读的次数。当一个新的数据包到来时，如果循环缓存区已满，则该数据包会被丢弃，从而造成数据包丢失。

用户循环缓存区大小比较重要，因为一方面它决定了在单个系统调用内，能够从内核空间复制到用户空间的最大数据量；另一方面，在单个系统调用中能够复制的最小数据量也非常重要。如果用户循环缓存区大小设置为一个较大的数值，则在从内核空间把数据复制到用户空间前，需要等待多个数据包的到来，降低了读取效率。NPF 具有可配置特性，允许用户在效率与好的响应度之间做出选择。Wpcap.dll 包括两个系统调用，能够用来设置读取数据超时的时间与内核层能够传输到应用程序的最小数据量。在默认状态下，读取数据超时的时间一般设置为 1s，内核层与应用程序之间，数据复制的最小数据量为 16 千字节。

2）数据包发送

NPF 允许将原始数据包发送到网络上。为发送数据，一个用户层应用程序在 NPF 设备上执行一个 WriteFile()系统调用。数据被发送到网络上，但不对数据进行任何协议封装，因此用户层应用程序将不得不构建每个数据包的不同协议头。用户层应用程序通常不需要生成 FCS，它通常由网络适配器计算，并在数据发送到网络之前，自动添加到一个数据包的尾部。在正常情况下，数据包的发送率并不是很高，因为每个数据包都需要一个系统调用，所以 Winpcap 添加了使用一次写系统调用就能把单个数据包发送多次的功能。通过 IOCTL 调用（控制码为 pBIOCSWRITEREP），用户层应用程序能够设置单个数据包发送的次数。例如，将发送次数设为 1000，用户层应用程序所写的每个原始数据包在驱动设备文件上都会发送 1000 次。在网络测试中，可利用此特性产生高速网络数据流：上下文切换的负载不再出现，因此性能显著提高了。

3）网络监视

Winpcap 提供的内核层的接口能够计算网络流量的简单统计信息。用户为了获得统计信息，不需要复制数据包到应用程序进行统计分析，只需要从统计引擎直接获取统计信息并显示。这避免了数据包捕获的大部分工作耗费在内存与 CPU 时钟上。

统计引擎由一个分类器和一个紧接分类器的计数器组成。数据包使用数据过滤器进行分类，这提供了一种配置方式来选择一个网络流量的子集。经过数据过滤器的数据传递到计数器，计数器保存一些变量，如数据包的数目与过滤器接收到字节的数量，并根据网络接收的数据包的数据来更新这些变量。这些变量在固定的时间间隔下传递给用户层应用程序，用户可以配置时间间隔的周期。采用此方法的优点是，在内核空间与用户空间中没分配缓存区。

4）转储到磁盘

Winpcap 提供将数据包从内核层转储到磁盘的功能，因此其可以在内核模式下将网络数据直接存储到磁盘空间。数据包捕获与内核层转储（libpcap 方式）如图 1-9 所示。

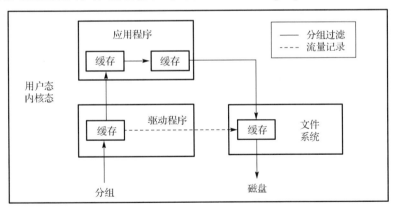

图 1-9　数据包捕获与内核层转储（libpcap 方式）

在图 1-9 中，实线箭头表示将数据包从内核层转储到磁盘的流程。在正常情况下，系统分配 4 个缓存区，以及捕获驱动程序缓存、应用程序存储捕获数据缓存、应用程序写文件时调用 stdio()函数（或类似）使用的缓存、文件系统使用缓存等。当 NPF 的内核层的网络数据流量记录特性被激活时，捕获驱动程序可直接访问文件系统。在图 1-9 中，虚线箭头表示可实现将捕获的数据包从内核层直接存储到磁盘。由此可见，在捕获数据包并存储的过程中，使用两个缓存区与一次复制是必须的，这样系统调用次数也少。当前，将数据包从内核层转储到磁盘的功能应用广泛，常使用上述 libpcap 方式，既提供了设置存入磁盘的数据包大小的功能，也提供了允许数据包在转储前被过滤的功能。

1.3.3　利用 winpcap 对数据包进行捕获和过滤

在共享式以太网环境下，当网络适配器设置为混杂模式时，由于共享式以太网采用广播信道争用的方式，因此监听系统可以捕获同一冲突域上传输的任意数据包。共享式以太网采用总线型结构，总线上每个站点网卡的主要工作是完成对总线当前状态的检测，以确定是否进行数据帧的传送，并判断每个物理数据帧的目的地址是否为本站地址。如果不匹配，则说明不是发送到本站的数据帧，需将其丢弃，否则接收该数据帧，并进行数据帧的 CRC 校验，然后将数据帧提交给上层 LLC 子层。连接在总线上的站点网卡具有以下 4 种工作模式。

（1）广播模式（Broadcast Model）：目的 MAC 地址是 0XFFFFFF 的数据帧为广播帧，工作在广播模式下的网卡仅可以接收二层广播帧。

（2）多播模式（Multicast Model）：以二层多播地址作为目的 MAC 地址的数据帧在被接收时，其仅可以被组内的其他主机同时接收，而组外主机则接收不到。如果将网卡设置为多播模式，则网卡可以接收所有的二层多播帧，而不论该帧是不是组内成员。

（3）直接模式（Direct Model）：工作在直接模式下的网卡只接收目的 MAC 地址是自己本地 MAC 地址的数据帧及所有广播帧。

（4）混杂模式（Promiscuous Model）：工作在混杂模式下的网卡接收所有流过网卡的数据帧。

网卡的默认工作模式包括广播模式和直接模式，即只接收广播帧和发给自己的数据帧。如果采用混杂模式，则一个站点网卡将接受同一网络内所有站点发送的数据帧，这样就可以达到对局域网内数据进行监视捕获的目的。NDIS 是微软和 3Com 公司联合制定的网络驱动规范，它提供了大量的操作函数接口，为上层的协议驱动提供服务，屏蔽了下层各种网卡的差别。NDIS 向上支持多种网络协议，如 TCP/IP、NWLink IPX/SPX、NETBEUI 等，向下支持不同厂家生产的多种网卡。NDIS 还支持多种工作模式和多种处理器，提供了一个完备的 NDIS 库（library），但库中提供的各函数都是工作在内核模式下的，用户不能直接调用操作。利用 Winpcap 实施数据包捕获和过滤的流程如下。

（1）打开网卡，设为混杂模式。

（2）Tap() 回调函数在得到监视命令后，从网络设备驱动程序处收集数据包并把监视到的数据包传送给过滤程序。

（3）当 NPF 监视到有数据包到达时，NDIS 的中间驱动程序先调用分组驱动程序，然后将数据包传递给每一个参与进程的分组过滤程序。

（4）由 NPF 决定哪些数据包应该丢弃、哪些数据包应该接收、是否需要将接收到的数据包复制到相应的应用程序。

（5）通过 NPF 过滤后，将未过滤掉的数据包提交给系统缓存区，待系统缓存区存满后，再将数据包复制到用户缓存区。

（6）监视程序直接从用户缓存区中读取捕获的数据包。

（7）关闭网卡。

1.3.4 主要数据结构和接口函数

编写基于 Winpcap 网络应用程序的第一步，就是获得本站点已连接的网络适配器设备列表，同时在程序结束时确保释放获取的网络适配器设备列表。

为了获得与释放已连接的网络适配器设备列表，Wpcap.dll 在文件/wpcap/libpcap/pcap/pcap.h 中提供了下列数据结构和函数说明。

```
struct pcap_if;          //网络适配器详细信息
struct pcap_addr;        //网络适配器地址信息
int pcap_findalldevs(pcap_if_t **alldevsp, char *errbuf);//获取网络适配器设备列表
void pcap_freealldevs(pcap_if_t *alldevsp ) ;        //释放网络适配器设备列表
```

在文件 wpcap/libpcap/remote-ext.h 中提供了对 int pcap_findalldevs()的功能扩展函数。

```
int pcap_findalldevs_ex(char *source, struct pcap_rmtauth *auth, pcap_if_t **alldevs, char*errbuf);
```

1）pcap_if 结构体

函数 pcap_findalldevs_ex()或 pcap_findalldevs()分别返回一个 pcap_if 类型的链表 alldevs 或 alldevsp。每个 pcap_if 结构体都包含一个网络适配器的详细信息，其中成员 name 和 description 分别表示一个网络适配器的名称和一个容易让人理解的描述。pcap_if 结构体的定义如下。

```
typedef struct pcap_if pcap_if_t;
struct pcap_if {
//如果不为 NULL，则指向链表的下一个元素；如果为 NULL，则为链表的尾部
struct pcap_if *next;
//pcap_open_live()函数传递一个描述网络适配器名称的字符串指针
char *name;
//如果不为 NULL，则指向描述网络适配器名称的一个可读字符串
char *description;
//一个指向接口地址链表的第一个元素的指针
struct pcap_addr *addresses;
//PCAP_IF_接口标志，当前仅有的可能标志为 PCAP_IF_LOOPBACK，如果接口是回环的，则设置该标志
bpf_u_int32 flags;
};
```

pcap_addr 结构体表示接口地址的信息，其定义如下。

```
typedef struct pcap_addr pcap_addr_t;
struct pcap_addr {
    struct pcap_addr *next;          /*指向下一个元素的指针*/
    struct sockaddr *addr;           /*IP 地址*/
    struct sockaddr *netmask;        /*网络掩码*/
```

```
    struct sockaddr *broadaddr;    /*广播地址*/
    struct sockaddr *dstaddr;      /*P2P目的地址*/
};
```

2）pcap_findalldevs_ex()函数

Winpcap 提供 pcap_findalldevs_ex()函数来获得已连接的网络适配器设备列表扩展功能，该函数的原型如下。

```
int pcap_findalldevs_ex(char *source, struct pcap_rmtauth *auth,
                        pcap_if_t **alldevs, char *errbuf);
```

用 pcap_findalldevs_ex()函数创建一个能用 pcap_open()函数打开的网络适配器设备列表，该函数是函数 pcap_findalldevs()的一个扩展。pcap_findalldevs()是一个过时的函数，其只允许列出本机上的网络设备；反之，pcap_findalldevs_ex()允许列出一个远程机器上的网络适配器设备，还能列出一个给定文件夹中可用的 pcap 文件，因为 pcap_findalldevs_ex()依赖于标准的 pcap_findalldevs()来获得本地机器的地址，支持平台无关性。

参数 source 可通知函数在哪儿查找设备的参数，并且它使用与 pcap_open()函数同样的语法。与 pcap_findalldevs()函数不同，设备的名称由 alldevs->name 指定。pcap_findalldevs()函数的输出必须采用 pcap_createsrcstr()函数进行格式处理，之后才能将源参数传递给 pcap_open()函数使用。

参数 auth 是一个指向 pcap_rmtauth 结构体的指针。该指针保存着 RPCAP 连接到远程主机上所需的认证信息。auth 参数对本地主机请求没什么意义，此时可以设为 NULL。

参数 alldevs 是一个 pcap_if 类型的指针，在函数成功返回时，该指针被设置为指向网络适配器设备列表的第一个元素，该列表的每个元素都是 pcap_if 类型的。

参数 errbuf 是一个指向用户分配的缓存区（大小为 PCAP_ERRBUF_SIZE）的指针，如果函数操作出现错误，则该缓存区将存储该错误的信息。

函数执行成功则返回 0，如果有错误则返回-1。如果变量 alldevs 返回网络适配器设备列表，当函数正确返回时，alldevs 不能为 NULL。也就是说，当系统没有任何接口时，该函数返回-1；变量 errbuf 返回错误信息。一个错误可能由下列原因导致。

（1）Winpcap 没有安装在本地/远程主机上。

（2）用户没有足够的权限来列出这些设备/文件。

（3）一个网络故障。

（4）RPCAP 版本协商失败（The RPCAP version negotiation failed）。

（5）其他错误（如没有足够的内存）。

3）pcap_findalldevs()函数

pcap_findalldevs()函数只允许列出本地主机上的网络适配器设备，该函数原型如下。

```
int pcap_findalldevs(pcap_if_t **alldevsp, char *errbuf);
```

pcap_findalldevs()函数获得已连接并能打开的网络适配器设备列表，该列表能够被 pcap_open_live()函数打开。

参数 alldevsp 指向网络适配器设备列表的第一个元素，该列表中的每个元素都为

pcap_if 类型。如果没有已连接并能打开的网络适配器设备，则该列表可能为 NULL。

如果该函数执行失败，则返回-1，参数 errbuf 存储当前的错误信息；如果该函数执行成功，则返回 0。

值得注意的是，通过调用 pcap_findalldevs()函数可能存在网络适配器设备不能被 pcap_open_live()函数打开的现象。例如，如果没有足够的权限来打开这些设备并进行捕获，则这些设备将不会出现在网络适配器设备列表中。

4）pcap_freealldevs()函数

通过 pcap_findalldevs_ex()函数或 pcap_findalldevs()函数获取网络适配器设备列表后，必须调用 pcap_freealldevs()函数释放该列表。pcap_freealldevs()函数的原型如下。

```
void pcap_freealldevs(pcap_if_t *alldevsp)
```

5）获得与释放网络适配器设备列表的实例

下列代码能获取网络适配器设备列表，并在屏幕上显示出来，如果没有找到网络适配器，则打印错误信息，并在程序结束时释放网络适配器设备列表。

```
#include "remote-ext.h"
#include "pcap.h"
main()
{
    pcap_if_t *alldevs;
    pcap_if_t *d;
    int i=0;
    char errbuf[PCAP_ERRBUF_SIZE];
    //获取本地网络适配器设备列表
    if (pcap_findalldevs_ex(PCAP_SRC_IF_STRING, NULL , &alldevs, errbuf) == -1)
    {
        //获取网络适配器设备列表失败，程序返回
        fprintf(stderr,"Error in pcap_findalldevs_ex: %s/n", errbuf);
        exit(1);
    }
    //获取网络适配器设备列表成功，打印该设备列表
    for(d= alldevs; d != NULL; d= d->next)
    {
        printf("%d. %s", ++i, d->name);
        if (d->description)
            printf(" (%s)/n", d->description);
        else
            printf(" (No description available)/n");
    }
    if (i = = 0)
    {
        //没找到设备接口，确认 Winpcap 已安装，程序退出
        printf("/nNo interfaces found!/n");
        Return 0;
```

```
        }
        //不再需要网络适配器设备列表了，释放它
        pcap_freealldevs(alldevs);
}
```

　　pcap_findalldevs_ex()函数和其他 libpcap()函数一样，有一个参数 errbuf，一旦函数执行发生错误，这个参数将会被 libpcap()写入字符串类型的错误信息中。如果在上面程序中遇到这种情况，会打印提示语句"No description available"。当完成了网络适配器设备列表的使用，需要调用 pcap_freealldevs()函数将其占用的内存资源释放出来。在某台 WinXP 的计算机上运行上述程序，得到的结果如下。

　　（1）/Device/NPF_{4E273621-5161-46C8-895A-48D0E52A0B83} (Realtek RTL8029(AS) Ethernet Adapter)。

　　（2）/Device/NPF_{5D24AE04-C486-4A96-83FB-8B5EC6C7F430} (3Com EtherLink PCI)。

　　当使用 Windows 系统提供的 Winpcap 库进行网络数据捕获时，首先要初始化两个结构体，一个是网络适配器的结构体 LpAdapter，一个是存放接收到的数据包的结构体 RecvPacket。

```
#define MAX_LINK_NAME_LENGTH 64
```

网络适配器数据结构如下。

```
typedef struct _ADAPTER
{
    HANDLE hFile;
    TCHAR Symboliclink[MAX_LINK_NAME_LENGTH];
    Int NumWrites;
} ADAPTER , *LPADAPTER;
```

　　说明：hFile 是一个指向网络适配器 HANDLE 的指针，通过它可以对网络适配器进行操作。Symboliclink 包含当前打开的网络适配器的名称。

　　数据包结构如下。

```
typedef struct _PACKET
{
    HANDLE      hEvent;
    OVERLAPPED  OverLapped;
    PVOID       Buffer;
    UINT        Length;
    PVOID       Next;
    UINT        ulBytesReceived;
    BOOLEAN     bIoComplete;     //控制接收数据包的开始和结束
} PACKET,*LPPACKET;
```

　　使用 Packet.dll 动态链接库编写程序，实现数据包捕获的步骤的主要源代码如下。

　　（1）获得网络适配器设备列表。

```
#define Max_Num_Adapter 10        //获得网络适配器设备列表中网络适配器的数量
char        Adapterlist[Max_Num_Adapter[512]] //获得网络适配器设备列表
int         i=0
```

```
char        AdapterNames[512],*tempa,*templa;
ULONG AdapterLength=1024;
```

（2）获得系统中网络适配器的名称。

```
PacketGetAdapterNames(AdapterNamea,&AdapterLength);
tempa=AdapterNamea;
templa=Adapternamea;
while ((*tempa!='/0')||(*tempa-1!='/0'))
{
    if (*tempa= ='/0')
    {
        memcpy(AdapterList[i],temla,tempa-templa);    //复制内存数据
        templa=tempa+1;
        i++
    }
    tempa++
}
```

（3）从网络适配器设备列表中选择一个默认为 0 号的网络适配器。

```
LPADAPTER lpAdapter
lpAdapter = PacketOpenAdapter (AdapterList[0]);
if (!lpAdapter||(lpAdapter->hFile==INVALID__HANDLE__VALUE))
{
    dwErrorCode=GetLastError();
    return FALSE;
}
```

（4）将选择的网络适配器 LpAdapter 设置为混杂模式。

`PacketSetHwFilter(LpAdapter,NDIS_PACKET_TYPE_PROMISCUOUS)`

（5）设置 BPF 内核层中包过滤的过滤器的参数。利用这个函数完成对原始数据包的初始过滤处理，如端口号、IP 地址等。

`PacketSetBpf (LpAdapter AdapterObject,structbpf_program*fp)`

（6）设置缓存区内存大小为 512 千字节。

`PacketSetBuff(lpAdapter, 512000);`

（7）分配一个数据包对象，并连接已分配的缓存区。

`PacketInitPacket(lpPacket, (char*)bufferReceive, 512000);`

（8）捕获多个数据包。网络适配器 LpAdapter 接收数据包，并将数据包放入 LpPacket 指向的数据包结构体中，若接收成功，则返回 TRUE，否则返回 FALSE。

`PacketReceivePacket(lpAdapter,lpPacket,TRUE);`

（9）通过触发回调函数，将捕获的符合过滤器规则的数据包转发给网络协议分析模块进行分析处理。

（10）停止接收数据包，释放数据包对象。

```
if(lpPacket!=NULL
{
    PacketFreePacket(lpPacket);
    lpPacket=NULL;
}
```

（11）关闭网络适配器，将网络适配器恢复到正常接收状态。

```
if(lpAdapter!=NULL)
{
    PacketCloseAdapter(lpAdapter);
    lpAdapter=NULL;
}
```

1.4 Winpcap 环境搭建

1.4.1 Visual Studio 10.0 实验环境搭建

（1）View→Property Manager→Debug|Win32 → Mircrosoft.Cpp.Win32.user（右键）→ Properties。

（2）设置环境目录：在 VC++ Directiories → Include Directories 和 Library Directories 中添加路径。

如果将 wpdpack 放到 c 盘，则 Include Directories: c:\wpdpack\Include;Library Directories: c:\wpdpack\Lib。

（3）在 Linker（连接器）下的 Command Line（命令行）Additional Options （附加项）中输入：wpcap.lib ws2_32.lib（用空格分隔）。

上述方法存在的问题：每次开机需要重新设置一次。为了实现一次设置一直有效，可采用如下方法。

（1）首先建立一个工程 networkstack。

（2）单击"文件"→"新建"→"项目"→"WIN 32 控制台应用程序"选项，输入项目名称（networkstack）和指定存储目录（D:\）；在"应用程序设置"对话框中，选择"应用程序类型"为"WIN 32 控制台应用程序"；选择"附加选项"为"空项目"；选择"添加公共头文件以用于"为"不选"。

（3）在"解决方案"标签中，选择"增加"或"产生新文件"选项。

（4）对环境进行配置，具体如下。

① 安装 WINPCAP.EXE。

② 将 WpdPack_4_1_2.zip 解压，并复制到 d:\根目录。

③ 在属性管理器标签中，单击创建的工程文件名，双击"debug|win32"，在弹出的"Debug 属性页"对话框中，双击"VC++目录"，在"包含目录"中，加入 WpdPack 复制的 INCLUDE 目录路径；在"库目录"中，加入 WpdPack 复制的 LIBARY 目录路径，如图 1-10 所示。

图 1-10　增加目录示意图

在图 1-10 的"Debug 属性页"对话框中，单击"链接器"→"输入"属性，打开以下页面（见图 1-11）。

在图 1-11 的"附加依赖项"中加入"wpcap.lib"和"ws2_32.lib"两个库文件。

图 1-11　增加库文件示意图

1.4.2　VC++6.0 环境配置

（1）下载 Winpcap 安装包，下载地址为：http://www.winpcap.org/install/default.htm。

点击 http://www.winpcap.org/devel.htm，下载 Winpcap developer's pack，然后解压，在获得的解压包中有配置好的例子和 include lib 和 example 等文件夹。

（2）在 VC++中设置 include 和 library 目录，具体方法如下。

① 打开 VC++，执行"TOOLS→option→directories"命令，在 include files 中添加…\wpdpack\include 目录（步骤 2 中得到的目录），在 library fillies 中添加…\wpdpack\lib 目录。

② 执行"project→settings→link"命令，在 object\library modules 中添加 wpcap.lib 和 Packet.lib 及 ws2_32.lib 3 个库文件。

第 2 章 工业以太网协议分析与实践

2.1 引　　言

网络上任意两个相邻节点之间都能够实现通信的前提是，节点与节点之间存在物理链路。该物理链路应采用有线或无线传输介质，同时必须有通信协议控制物理链路的数据传输；若把实现协议的硬件和软件附加到通信物理链路上，就构成了数据链路；要在技术上解决信号传输的可靠性比较难，但如果通过软件或硬件，以通信协议的形式实现信号（帧）传输的可靠性则相对比较容易。因此，设计数据链路层的目的是通过通信协议将不可靠的物理链路转变为可靠的数据链路。

由此可见，数据链路（实际上是一条逻辑链路） = 物理链路 + 数据链路层协议。在通信网络中，一般使用网络适配器（网卡或网络接口）实现数据链路层协议。一般网络适配器包括物理层和数据链路层。数据链路层协议的发展经历了从面向字符型到面向比特型的过程；典型的面向字符型的通信规程有 IBM 公司的二进制同步通信（Binary Synchronous Communication，BSC）规程和国际标准化组织（ISO）的基本型控制规程（ISO-1745 标准）。由于这种通信规程与特定的字符编码集的关系过于密切，兼容性较差，并且在实现上也比较复杂，因此在现代的数据通信系统中已很少使用。面向比特型的通信规程是由 IBM 公司于 20 世纪 70 年代初提出的，称为同步数据链路控制（Synchronous Data Link Control，SDLC）规程。美国国家标准学会（ANSI）和 ISO 在此基础上进行了规范和发展，分别形成了各自的标准，即 ANSI 的先进数据通信控制规程（Advanced Data Communication Control Procedure，ADCCP），以及 ISO 的高级数据链路控制（High level Data Link Control，HDLC）规程。此外，在 ITU X.25 建议中，链路级采用了一种 HDLC 的变种，称为链路访问规程（Link Access Procedure，LAP）或平衡链路访问规程（Link Access Procedure Balanced，LAPB），并以 LAPB 为主要通信模式。数据链路层 LLC 子层通信协议源于 HDLC。

上述通信规程尽管在一些细节上存在差异，但从实现功能上来说大同小异。

2.2 工业以太网协议工作原理

为了简化相邻两个网络节点之间的通信过程，工业以太网在数据链路层采取了两个重要的措施。

（1）由于现代网络传输介质采用双绞线和光纤，误码率低，因此数据链路层 LLC 子层不再提供差错控制和流量控制。

（2）由于数据链路层不需要提供数据帧传输的可靠性要求，因此工业以太网对发送的数据帧不再进行编号，也不要求对方发回确认（DIX Ethernet V2），理由是局域网信道的质量很好，所以因信道质量产生比特差错的概率很小。

由此可见，实际上工业以太网提供的服务是相邻两个网络节点之间不可靠的通信，即尽最大努力传输数据帧。当目的站总收到有比特差错的数据帧时就丢弃该帧，其他什么也不做；差错的纠正由高层（传输层 TCP 协议）决定。当高层发现丢失了一些数据帧而启动重传机制时，工业以太网并不知道这是一个重传的数据帧，而是将其当作一个新的数据帧来发送。

2.2.1 工业以太网协议语法

工业以太网协议语法指发送方产生的数据帧格式。数据帧包括 6 个字段：目的 MAC 地址（6 字节）字段、源 MAC 地址（4 字节）字段、协议类型（4 字节）字段、数据字段、填充字段及 FCS 校验（4 字节）字段。物理层通过硬件在数据帧前面加 7 个字节的前导码和 1 个字节的帧定界符，如图 2-1 所示。

图 2-1 工业以太网数据帧格式（协议语法）

1．前导码（PA）

帧同步序列的前导码格式为连续 7 个字节的"10101010"二进制序列。该序列经过曼彻斯特编码之后，会产生一个 10MHz 的方波，使接收节点的时钟和发送方的时钟同步。前导码字段不作为数据帧的有效部分。

2．帧定界符（SFD）

帧定界符表示一个有效数据帧的开始，其格式为"10101011"二进制序列。帧定界符字段不作为数据帧的有效部分。

3．目的 MAC 地址（DA），源 MAC 地址（SA）

一个完整数据帧开始于目的 MAC 地址字段，结束于冗余校验字段。MAC 地址也称硬件地址。当目的 MAC 地址的最高位为"0"时，表示二层单播地址，只能代表一个网络接口；当目的 MAC 地址的最高位为"1"时，表示组播地址，代表一组网络接口；当目的 MAC 地址全为"1"时，表示广播地址。目的 MAC 地址的类型可以是二层单播地址、组播地址或广播地址，但源 MAC 地址只能是单播地址。目的 MAC 地址由 IEEE 分配和管理，原则上具有全球唯一性。目的 MAC 地址的前 3 个字节标识厂商，如 CISCO 公司的标识为 00-00-0C；IBM 公司的标识为 08-00-5A；HP 公司的标识为 08-00-09 等。目的 MAC 地址的后 3 个字节由网络接口制造厂商独立分配与管理。实际上，目的 MAC 地址只需要在局域网内唯一，不需要全球唯一，因为目的 MAC 地址的使用范围只在一个网段内有效，而不同网段之间的网络是通过基于 IP 地址路由转发实现的，与目的 MAC 地址没有关系。

4．协议类型（TYPE）

协议类型表示上层协议类型，接收方利用该字段将数据帧中的数据部分交付给上层协议。如果发送方是 IP 协议调用链路层协议接口，则该协议字段为 0X0800；接收方根据协议类型，将数据帧的数据部分交付给上层 IP 协议进行处理。0X0800 表示上层 IP 协议，0X8137 表示 IPX 协议，0x0806 表示 ARP 协议。

5．数据（DATA）

数据表示要传送的网络层协议数据单元，是数据帧的数据部分。网络层协议数据单元的数据长度应是字节的倍数，最大为 1500 字节，最小为 46 字节；如果数据长度小于 46 字节，则发送方需要填充若干 0 以达到 46 字节；接收方数据链路层无法知道填充了多少个 0，只能依靠高层网络层 IP 协议或 ARP 协议的语法及语义规则来判断。例如，接收方在处理 IP 分组时，IP 分组首部中有一个总长度字段，表示完整 IP 分组有多大，这样可以区分填充字段并丢弃。

6．填充（PAD）

工业以太网数据链路层数据帧的数据部分的长度有 MTU 要求：大于 46 字节，不小于 1500 字节。因此，数据帧最大长度为 1518 字节，最小长度为 64 字节。如果数据帧协议数据单元（PDU）与填充字段长度小于 46 个字节，则必须在填充字段上填充若干字节的 0，使 PDU 和填充字段的总长度不小于 46 个字节；否则，接收节点会把该超短帧作为"帧碎片"过滤掉，不予接收，同时要确保帧长为字节的整数倍，如图 2-2 所示。

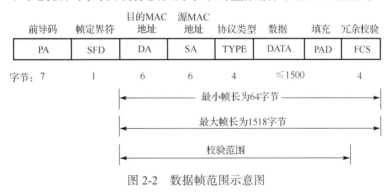

图 2-2　数据帧范围示意图

7．冗余校验（FCS）

数据帧采用 32 位 CRC 校验码（CRC-32），校验范围包括目的 MAC 地址+源 MAC 地址+协议类型+数据+填充等字段。使用生成多项式的方式计算校验码，具体如下。

$$G(x)-32 = x^{32} + x^{26} + x^{23} + x^{22} + x^{16} + x^{12} + x^{11} + x^{10} + x^8 + x^7 + x^5 + x^4 + x^2 + x + 1$$

2.2.2 工业以太网协议语义

工业以太网数据链路层发送方通常可构建和发送广播帧和单播帧。

广播帧：目的 MAC 地址为 48 个 1：0XFFFFFFFFFFFF，表示为二层广播帧。广播帧不能穿透三层设备，只能限制在本地网段内传输；在交换式网络中，广播帧可以被网段内的所有网络接口接收到。

单播帧：如果目的 MAC 地址的第一个比特位为 0，则表示一个单播帧；在交换式网络中，单播帧只能被网段内的某一个网络接口按照"目的 MAC 地址匹配"原则接收到。

2.2.3 工业以太网协议时序关系

发送方时序关系如下。
（1）定义工业以太网数据帧格式数据结构。
（2）组装数据帧首部。
（3）组装数据帧数据部分。
（4）计算校验码（CRC-32）。
（5）调用 Winpcap 接口函数，发送数据帧。

接收方时序关系如下。
（1）定义工业以太网数据帧格式数据结构。
（2）调用 Winpcap 接口函数，选择当前活动网卡号。
（3）接收数据帧。
（4）对接收到的数据帧进行基本处理：如果目的 MAC 地址是二层广播地址或本地 MAC 地址，则接收，否则丢弃；如果数据帧长度在 64～1518 字节，则接收，否则丢弃；如果校验码（CRC-32）检查正确，则接收，否则丢弃。

2.3 工业以太网发送数据帧

工业以太网数据链路层发送方（源节点）实现源代码以下。

```
include<stdio.h>
#include<stdlib.h>

#define HAVE_REMOTE
#include<pcap.h>

#pragma warning(disable:4996)
#define ETHERNET_IP 0x0800
#define MAX_SIZE 2048
int size_of_packet = 0;
u_int32_t crc32_table[256];         //存储CRC-32计算表
struct ethernet_header              //工业以太网首部
{
```

```c
    u_int8_t dest_mac[6];           //目的 MAC 地址
    u_int8_t src_mac[6];            //源 MAC 地址
    u_int16_t ethernet_type;        //协议类型
};
void generate_crc32_table()         //产生 CRC-32 计算表 (generate table)
{
    int i, j;
    u_int32_t crc;
    for (i = 0; i < 256; i++)
    {
        crc = i;
        for (j = 0; j < 8; j++)
        {
            if (crc & 1)
                crc = (crc >> 1) ^ 0xEDB88320;
            else
                crc >>= 1;
        }
        crc32_table[i] = crc;
    }
}
//利用查表法计算 4 字节 CRC 校验码
u_int32_t calculate_crc(u_int8_t *buffer, int len)
{
    int i, j;
    u_int32_t crc;
    crc = 0xffffffff;
    for (i = 0; i < len; i++)
    {
        crc = (crc >> 8) ^ crc32_table[(crc & 0xFF) ^ buffer[i]];
    }
    crc ^= 0xffffffff;
    return crc;
}
//组装数据帧首部
void load_ethernet_header(u_int8_t *buffer)
{
    struct ethernet_header *hdr = (struct ethernet_header*)buffer;
    //填写目的 MAC 地址
    hdr->dest_mac[0] = 0x00;
    hdr->dest_mac[1] = 0x09;
    hdr->dest_mac[2] = 0x73;
    hdr->dest_mac[3] = 0x07;
    hdr->dest_mac[4] = 0x74;
    hdr->dest_mac[5] = 0x73;
```

```c
    //填写源 MAC 地址
    hdr->src_mac[0] = 0x00;
    hdr->src_mac[1] = 0x09;
    hdr->src_mac[2] = 0x73;
    hdr->src_mac[3] = 0x07;
    hdr->src_mac[4] = 0x73;
    hdr->src_mac[5] = 0xf9;
//填写协议类型
    hdr->ethernet_type = ETHERNET_IP;
    size_of_packet += sizeof(ethernet_header);
}
//组装从文件中读取的数据作为数据帧的数据部分
int load_ethernet_data(u_int8_t *buffer, FILE *fp)
{
    int size_of_data = 0;
    char tmp[MAX_SIZE], ch;
    while ((ch = fgetc(fp)) != EOF)
    {
        tmp[size_of_data] = ch;
        size_of_data++;
    }
    if (size_of_data < 46 || size_of_data>1500)
    {
        printf("Size of data is not satisfied with condition!!!\n");
        return -1;
    }
    //计算得到 4 字节 CRC 校验码
    u_int32_t crc = calculate_crc((u_int8_t*)tmp, size_of_data);
    //填写 CRC 校验码
    for (i = 0; i < size_of_data; i++)
    {
        *(buffer + i) = tmp[i];
    }
    *(u_int32_t*)(buffer + i) = crc;
    size_of_packet += size_of_data + 4;
    return 1;
}
int main()
{
    //存储要发送的数据帧的缓存
    u_int8_t buffer[MAX_SIZE];
    generate_crc32_table();
    //创建一个数据帧
    size_of_packet = 0;
    FILE *fp = fopen("data.txt", "r");
    if (load_ethernet_data(buffer + sizeof(ethernet_header), fp) == -1)
```

```
    {
        return -1;
    }
    load_ethernet_header(buffer);
    //发送数据帧
    pcap_t *handle;
    char *device;
    char error_buffer[PCAP_ERRBUF_SIZE];
    device = pcap_lookupdev(error_buffer);
    if (device == NULL)
    {
        printf("%s\n", error_buffer);
        return -1;
    }
    //选择 1 号网络接口发送数据帧
    handle = pcap_open_live(device, size_of_packet, PCAP_OPENFLAG_PROMISCUOUS, 1,
        error_buffer);
    if (handle == NULL)
    {
        printf("Open adapter is failed..\n");
        return -1;
    }
    //发送 20 个不同数据帧
    int i = 20;
    while(i--)
    {
        pcap_sendpacket(handle, (const u_char*)buffer, size_of_packet);
        printf("%d", *(int*)(buffer + size_of_packet - 4));
    }
    //释放数据链路层资源
    pcap_close(handle);
    return 0;
}
```

2.4 工业以太网接收数据帧

工业以太网数据链路层接收方（目的节点）实现源代码如下。

```
#include<stdio.h>
#include<stdlib.h>
#define HAVE_REMOTE
#include<pcap.h>
#include<WinSock2.h>
#pragma warning(disable:4996)
//对接收数据帧进行处理的回调函数
void ethernet_protocol_packet_callback(u_char *argument, const struct
```

```c
                pcap_pkthdr *packet_header,
    const u_char *packet_content);
//数据帧首部前3个字段数据结构
struct ethernet_header
{
    u_int8_t ether_dhost[6];            //目的MAC地址
    u_int8_t ether_shost[6];            //源MAC地址
    u_int16_t ether_type;               //协议类型
};
//可接收的数据帧的目的MAC地址初始化
u_int8_t accept_dest_mac[2][6] = { { 0x11, 0x11, 0x11, 0x11, 0x11, 0x11 },
            { 0x00, 0x09, 0x73, 0x07, 0x74, 0x73 } };
//存储计算校验码（CRC-32）的表
u_int32_t crc32_table[256];
//产生计算校验码（CRC-32）的表的内容
void generate_crc32_table()
{
    int i, j;
    u_int32_t crc;
    for (i = 0; i < 256; i++)
    {
        crc = i;
        for (j = 0; j < 8; j++)
        {
            if (crc & 1)
                crc = (crc >> 1) ^ 0xEDB88320;
            else
                crc >>= 1;
        }
        crc32_table[i] = crc;
    }
}
//计算校验码（CRC-32）
u_int32_t calculate_crc(u_int8_t *buffer, int len)
{
    int i, j;
    u_int32_t crc;
    crc = 0xffffffff;
    for (i = 0; i < len; i++)
    {
        crc = (crc >> 8) ^ crc32_table[(crc & 0xFF) ^ buffer[i]];
    }
    crc ^= 0xffffffff;
    return crc;
}
```

```c
//对接收到的数据帧进行分析
void ethernet_protocol_packet_callback(u_char *argument, const struct
    pcap_pkthdr *packet_header, const u_char *packet_content)
{
    u_short ethernet_type;
    struct ethernet_header *ethernet_protocol;
    u_char *mac_string;
    static int packet_number = 1;
    ethernet_protocol = (struct ethernet_header*)packet_content;
    //获得数据帧前3个字段
    int len = packet_header->len;
    int i, j;
    //如果接收到的数据帧的目的MAC地址与本地MAC地址或二层广播地址匹配则接收, 否则丢弃
    int flag = 2;
    for (i = 0; i < 2; i++)
    {
        flag = 2;
        for (j = 0; j < 6; j++)
        {
            if (ethernet_protocol->ether_dhost[j] == accept_dest_mac[i][j])
                continue;
            else
            {
                flag = i;
                break;
            }
        }
        if (flag != 2)continue;
        else
            break;
    }
    if (flag != 2)
    {
        return;
    }
    if (i == 0)
    {
        printf("It's broadcasted.\n");
    }
    //检测校验码(CRC-32), 如果正确则接收, 否则丢弃
    u_int32_t crc = calculate_crc((u_int8_t*)(packet_content + sizeof
        (ethernet_header)), len - 4 - sizeof(ethernet_header));
    if (crc != *((u_int32_t*)(packet_content + len - 4)))
    {
        printf("The data has been changed.\n");
        return;
```

```c
        }
        printf("----------------------------\n");
        //打印接收数据帧的数量
        printf("capture %d packet\n", packet_number);
        //打印接收数据帧的时间
        printf("capture time: %d\n", packet_header->ts.tv_sec);
        //打印数据帧的首部长度
        printf("packet length: %d\n", packet_header->len);
        printf("-----Ethernet protocol-------\n");
        //打印数据帧协议类型字段的值
        ethernet_type = ethernet_protocol->ether_type;
        printf("Ethernet type: %04x\n", ethernet_type);
        switch (ethernet_type)
        {
            case 0x0800:printf("Upper layer protocol: IPv4\n"); break;
            case 0x0806:printf("Upper layer protocol: ARP\n"); break;
            case 0x8035:printf("Upper layer protocol: RARP\n"); break;
            case 0x814c:printf("Upper layer protocol: SNMP\n"); break;
            case 0x8137:printf("Upper layer protocol: IPX\n"); break;
            case 0x86dd:printf("Upper layer protocol: IPv6\n"); break;
            case 0x880b:printf("Upper layer protocol: PPP\n"); break;
            default:
                break;
        }
        //打印数据帧的源 MAC 地址
        mac_string = ethernet_protocol->ether_shost;
        printf("MAC source address: %02x:%02x:%02x:%02x:%02x:%02x\n",
            *mac_string, *(mac_string + 1), *(mac_string + 2), *(mac_string
            + 3),*(mac_string + 4), *(mac_string + 5));
        //打印数据帧的目的 MAC 地址
        mac_string = ethernet_protocol->ether_dhost;
        printf("MAC destination address: %02x:%02x:%02x:%02x:%02x:%02x\n",
           *mac_string, *(mac_string + 1), *(mac_string + 2),
           *(mac_string + 3), *(mac_string + 4), *(mac_string + 5));
        /*if (ethernet_type == 0x0800)
        {
            ip_protocol_packet_callback(argument, packet_header, packet_content +
                sizeof(ethernet_header));
        }
        */
        //打印数据帧数据部分的内容
        for (u_int8_t *p = (u_int8_t*)(packet_content + sizeof(ethernet_header)); p !=
            (u_int8_t*)(packet_content + packet_header->len - 4); p++)
        {
            printf("%c", *p);
        }
```

```c
    printf("\n");
    printf("-----------------------\n");
    packet_number++;
}

int main()
{
    //产生计算校验码（CRC-32）的表
    generate_crc32_table();
    pcap_if_t *all_adapters;
    pcap_if_t *adapter;
    pcap_t *adapter_handle;
    char error_buffer[PCAP_ERRBUF_SIZE];
    //获取接收方计算机设备（网络接口）列表
    if (pcap_findalldevs_ex(PCAP_SRC_IF_STRING, NULL, &all_adapters,
        error_buffer) == -1)
    {
        fprintf(stderr, "Error in findalldevs_ex function: %s\n", error_buffer);
        return -1;
    }
    if (all_adapters == NULL)
    {
        printf("\nNo adapters found! Make sure WinPcap is installed!!!\n");
        return 0;
    }
    //打印接收方计算机设备列表中所有网络接口的名称和描述
    int id = 1;
    for (adapter = all_adapters; adapter != NULL; adapter = adapter->next)
    {
        printf("\n%d.%s\n", id++, adapter->name);
        printf("--- %s\n", adapter->description);
    }
    printf("\n");

    int adapter_id;
    //用户选择接收数据帧用到的网络接口
    printf("Enter the adapter id between 1 and %d: ", id - 1);
    scanf("%d", &adapter_id);
    if (adapter_id<1 || adapter_id>id - 1)
    {
        printf("\n Adapter id out of range.\n");
        pcap_freealldevs(all_adapters);
        return -1;
    }

    adapter = all_adapters;
```

```c
    for (id = 1; id < adapter_id; id++)
    {
        adapter = adapter->next;
    }
    //打开用户选择的网络接口
    adapter_handle = pcap_open(adapter->name, 65535, PCAP_OPENFLAG_PROMISCUOUS, 5,
            NULL, error_buffer);
    if (adapter_handle == NULL)
    {
        fprintf(stderr, "\n Unable to open adapter: %s\n", adapter->name);
        pcap_freealldevs(all_adapters);
        return -1;
    }
    //发送数据帧
    pcap_loop(adapter_handle, NULL, ethernet_protocol_packet_callback, NULL);
    //释放数据链路层资源
    pcap_close(adapter_handle);
    pcap_freealldevs(all_adapters);
    return 0;
}
```

第 3 章　共享式以太网协议分析与实践

3.1 引　　言

　　局域网（LAN）是指传输距离有限、传输速率高、以共享网络资源为主要目的的网络。常见的局域网主要有以太网、令牌总线网、令牌环网、无线局域网（WLAN）、光纤分布式数据接口（FDDI）、100VG-ANYLan 网络，其中以太网、令牌总线网、令牌环网、WLAN 等都是 IEEE 标准，FDDI 是 ANSI 标准。随着局域网市场的竞争与发展，目前以太网和 WLAN 占主导地位。IEEE 802 委员会于 1980 年 2 月成立，主要从事局域网标准化方面的工作，目的是推动局域网技术的应用，规范局域网产品开发的标准。IEEE 802 委员会主要由 3 个部门组成，具体如下。

　　（1）通信介质分会：研究 OSI 参考模型物理层，涉及局域网物理层信号传输特性、传输介质和物理接口。

　　（2）信号存取控制分会：研究 OSI 参考模型数据链路层，主要涉及逻辑链路控制协议和介质访问控制协议，以及这两个协议子层间的接口。

　　（3）高层接口分会：研究局域网对 OSI 参考模型高层的影响，具体包括局域网标准对 OSI 参考模型从网络层到应用层的影响。

　　IEEE 802 委员会制定的局域网标准是一个标准系列，并随着局域网技术的发展而不断增加新的标准，如图 3-1 所示。IEEE 802 系列标准主要规定了物理层和数据链路层两个层次，数据链路层分为逻辑链路控制（LLC）和介质访问控制（MAC）两个子层。LLC 子层与传输介质、物理接口无关，因此不管采用何种协议局域网，其对 LLC 子层来说都是透明的，主要解决数据帧在两个相邻站点之间传输的可靠性，如图 3-2 所示。局域网中与物理接口、传输介质、介质访问控制有关的数据链路层内容都放在 MAC 子层实现，所有局域网在 LLC 子层遵循统一规范：IEEE802.2 标准。所以通过 LLC 子层屏蔽了所有不同局域网异构性，如图 3-3 所示。IEEE 802 系列标准还包含控制网际互联部分（IEEE802.1 标准），其通过协议确保不同的局域网之间、局域网和广域网之间的兼容性。

　　不同的局域网主要指物理接口和传输介质不同。传输介质不同，MAC 子层协议也不同，但任意一个局域网 LLC 子层均遵循 IEEE 802.2 标准。根据 IEEE 802 委员会观点，从网络层上看，所有的局域网都是相同的。

　　（1）IEEE 802.1 标准：体系结构及其互连。

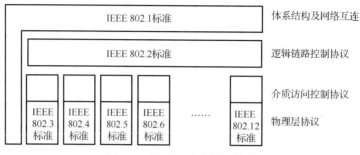

图 3-1 局域网标准间的关系

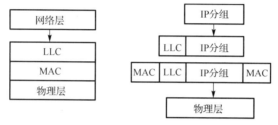

图 3-2 局域网分层结构和数据封装

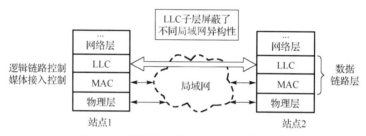

图 3-3 局域网相邻站点之间的通信方式

（2）IEEE 802.2 标准：LLC 子层规范。

（3）IEEE 802.3 标准：以太网 CSMA/CD 协议。

（4）IEEE 802.4 标准：令牌总线网规范。

（5）IEEE 802.5 标准：令牌环网规范。

（6）IEEE 802.8 标准：FDDI 网规范。

（7）IEEE 802.11 标准：无线局域网规范。

（8）IEEE 802.15 标准：无线个域网规范。

（9）IEEE 802.3 标准到 IEEE 803.12 标准定义了相应局域网的 MAC 子层和物理层标准，各网络在 LLC 子层遵循相同的 IEEE 802.2 标准。

网络接口（见图 3-4）又称网络适配器或网卡（NIC），计算机上的网络接口一般称为网卡，网络交换设备上的接口称为接口模块。网络适配器的重要功能如下。

（1）实现以太网数据链路层协议（LLC+MAC）及物理层功能。

（2）对发送和接收的数据帧进行缓存。

（3）进行串行/并行转换。

（4）接受操作系统中安装的网卡驱动程序管理、监控、参数配置。

一般有这个的都有集成网卡

图 3-4　网络接口（网卡）结构示意图

以太网最初是由美国施乐（Xerox）公司和斯坦福大学联合开发的。1975 年，相关网络产品上市。DEC、Intel、Xerox 等公司于 1980 年 9 月第一次公布了以太网的物理层和数据链路层标准，其成为世界上第一个局域网工业标准（以太网标准 V1 版本）；1982 年对该标准进行了修订，目前共享式以太网还在使用（以太网标准 V2 版本）。

20 世纪 80 年代，由于星型拓扑和结构化布线的 10BASE-T 以太网出现，因此共享式以太网的性能得到大大增强。共享式以太网采用半双工通信模式，主要特征是 MAC 子层使用 CSMA/CD 协议，交换设备主要为中继器和集线器。

20 世纪 90 年代，以太网交换机的问世标志着全双工以太网及快速交换技术的出现，使以太网在局域网中占据了主流。该类型网络称为交换式以太网，一般采用全双工通信模式，MAC 子层不再使用 CSMA/CD 协议，而是采用基于 MAC 地址转发表来完成数据帧的转发。1998 年，千兆位以太网技术出现，确立了以太网在局域网中的霸主地位。1983 年，IEEE 802.3 标准（MAC 子层标准+物理层标准）的公布，明确了数据链路层采用 LLC+MAC 两个子层。2002 年，IEEE 802 委员会批准组建了 10G 城域以太网（MAN），并成立了城域网论坛，主要讨论以太网用于城域网需要解决的问题。为将以太网技术由局域网上升为城域网，甚至上升为广域网进行技术准备。2007 年 7 月，IEEE 802 委员会批准了研究 40G/100G 下一代以太网标准的计划。目前，以太网按其传输速率可分为以下 3 种。

（1）标准以太网：10MB/s 以太网。

（2）快速以太网：100MB/s 以太网。

（3）高速以太网：1000MB/s 以上以太网。

每种以太网根据不同的传输介质、通信操作方式，以及采用不同 MAC 子层标准与物理层标准，形成了 IEEE 802.3 标准系列。

（1）IEEE 802.3i：10BASE-T（T：Twist pair）MAC 子层标准与物理层标准。

（2）IEEE 802.3u：100BASE-T MAC 子层标准与物理层标准。

（3）IEEE 802.3ab：1000MB/s 以太网，MAC 子层标准与物理层标准（半双工通信模式）。

（4）IEEE 802.3z：1000MB/s 以太网，MAC 子层标准与物理层标准（全双工通信模式）。

（5）IEEE 802.3ae：10GB/s 以太网，MAC 子层标准与物理层标准。

根据传输介质访问方式不同，以太网可分为共享式以太网和交换式以太网。共享式以太网采用半双工通信模式，总线型结构特征，网络连接设备为中继器和集线器，如标准以太网和部分快速和高速以太网，MAC 子层采用 IEEE 802.3 标准。交换式太网采用全双工通信模式，网络连接设备为网桥和二层/三层交换机，如部分快速和高速以太网，但 MAC 子层不再使用 IEEE 802.3 标准，而在 MAC 子层采用基于 MAC 地址转发表来完成数据帧的转发。标准以太网一定为共享式以太网，不论其采用什么传输介质，MAC 子层遵循 IEEE 802.3 标准，只是根据不同网络物理层而有所变化，主要网络连接设备为中继器和集线器。快速以太网不论采用什么传输介质，根据通信模式分为共享式以太网和交换式以太网。高速以太网同快速以太网相似，基本全为交换式以太网，网络交换设备为二层/三层交换机。

共享式以太网是将多个计算机连接到一根无源总线上，当初认为这样的连接方法简单，因为总线上没有有源器件。共享式以太网发送的数据帧在物理层都使用曼彻斯特编码或差分曼彻斯特编码，如图 3-5 所示。曼彻斯特编码的优点是，采用自同步编码方式，在通信时只需要一根数据线，不需要同步线，但缺点是信号速率是数据速率的两倍。

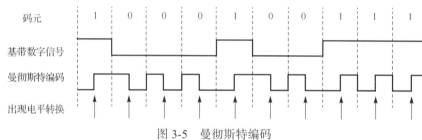

图 3-5 曼彻斯特编码

IEEE 802.3 标准是在以太网标准的基础上制定的。目前，以太网按其传输速率可分成 10MB/s 以太网、100MB/s 以太网和 1000MB/s 以太网，每种以太网根据不同的传输介质有多种物理子标准，形成了 IEEE 802.3 标准系列。无论何种以太网，其 MAC 子层均采用争用型介质访问控制协议，其核心思想是载波帧听、多路访问/冲突检测（CSMA/CD 协议）。共享式以太网组网灵活简便，传输介质一般采用同轴电缆。

3.2 共享式以太网的工作原理

共享式以太网的 MAC 子层主要定义了争用型介质访问控制协议，以及数据帧的封装与发送、数据帧的接收与解封等功能。共享式以太网采用的通信协议为 CSMA/CD 协议，该协议是一种争用型介质访问控制协议。CSMA/CD 协议在美国夏威夷大学开发的 ALOHA 网络系统采用的 ALOHA 协议的基础上进行了改进，提高了介质利用率。CSMA/CD 协议是一种分布式介质访问控制协议，网络中的各站点都能独立地决定数据帧的发送与接收。在一条总线上可连接多个网络站点，每个站点依据争用型介质访问控制协议可自主决定是否发送和接收数据帧。每个站点在发送数据帧之前，先侦听总线是否有载波，以判断总线是否空闲，只有总线处于空闲状态时，才允许发送数据帧。如果两个以上的站点同时监听到总线空闲并发送数据帧，那么会产生冲突现象，导致数据帧受到损坏，从而成为无效的数据帧。被损坏的数据帧必须重新发送。每个站点都有能力随时检测冲突是否发生，一旦发生冲突，则应停止发送数据帧，以免介质带宽因传送无效帧而白白浪费。然后随机延时一段时间，再重新争用介质重发数据帧。CSMA/CD 协议简单、可靠，采用该协议的共享式以太网在某段时间内被广泛使用。

3.2.1 共享式以太网数据帧语法及语义

IEEE 802.3 标准的 CSMA/CD 协议定义了数据帧格式，如图 3-6 所示。

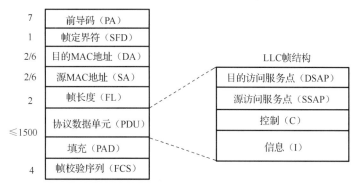

图 3-6 CSMA/CD 协议的数据帧格式

数据帧格式中各字段的含义如下。

（1）前导码（PA）：帧同步序列，其格式为连续 7 个字节的"10101010"二进制序列；它的作用是使接收站点的接收电路在正式开始接收帧之前达到稳定的同步状态，但它不作为帧的有效部分。

（2）帧定界符（SFD）：表示一个有效帧的开始，其格式为"10101011"二进制序列，它也不作为帧的有效部分。

（3）目的 MAC 地址（DA），源 MAC 地址（SA）：分别表示目的站点和源站点（发送站点）地址，可以选择 16 位或 48 位地址长度，但这两个地址长度必须保持一致。目的 MAC 地址可以是二层单播地址、多播地址或广播地址；而源 MAC 地址必须是单播地址。在选用 48 位地址长度时，可用特征位来指示该地址是作为局部地址的，还是作为全局地址。

（4）帧长度（FL）：以字节为单位来表示协议数据单元的实际长度。

（5）协议数据单元（PDU）：表示要传送的 LLC 子层数据。LLC 子层数据应是一个字节序列，最大数据长度为 1500 个字节。

（6）填充（PAD）：数据帧要求有最小帧长限制，最小帧长为 64 个字节，其中包括 18 个字节固定长度的帧头（帧头为目的 MAC 地址、源 MAC 地址、帧长度和冗余校验等 4 个字段，共 18 个字节）在内。如果实际的协议数据单元数据长度小于 46 个字节，则必须在填充字段上填充若干字节的 0，使协议数据单元字段和填充字段的总长度不小于 46 个字节；否则，接收站点会把超短帧作为"帧碎片"过滤掉，不予接收。

（7）帧校验序列（FCS）：采用 32 位 CRC 校验码，用规定的生成多项式的方式去除数据信息，获得的余数作为校验序列填入 FCS 字段。

因此，包括 18 个字节的帧头和帧尾在内的最大帧长为 1518 个字节。

从图 3-6 中数据帧的结构可以看出，MAC 子层协议产生的协议数据单元在 LLC 子层协议数据单元的外面，加上帧头和帧尾，组装成一个完整的数据帧，再经物理层编码传送出去的。数据链路层帧的封装过程如下。

（1）上层的信息 I 在经过 LLC 子层时被封装成 LLC 帧。其中，DSAP、SSAP 是服务访问点地址，是一种逻辑接口，以便在源站点和目的站点的对等协议层之间建立通信关系，目的站点将接收的信息 I 提交给 DSAP 指示的上层协议。

（2）当 LLC 子层经过 MAC 子层时又被封装成数据帧。其中，目的 MAC 地址、源 MAC 地址是目的站点地址和源站点地址，主要在两个站点之间建立通信关系，站点将根据目的 MAC 地址来确定是否接收数据帧。如果站点地址与目的 MAC 地址相匹配，则接收该数据帧；否则，不接收该数据帧。由此可见，数据帧必须通过这样的层层封装，才能实现数据传输。

目的站点要对接收到的数据帧进行拆封，其拆封过程与封装过程正好相反，即一层一层地去掉附加的地址信息和辅助信息，最后只将信息 I 提交给由 DSAP 指示的上层协议。

3.2.2 共享式以太网数据帧的发送及其过程分析

共享式以太网 MAC 子层协议遵循 IEEE 802.3 标准，其核心是 CSMA/CD 协议。该标准的工作原理类似现实生活中在楼道中喊话的过程，由此可见，新技术源于生活，技术创新就在身边。实际上，CSMA/CD 协议的产生并不是一蹴而就的，其也有一个发展的历史过程，如图 3-7 所示。

图 3-7　CSMA/CD 协议的发展过程

共享式以太网的基本特征是，各计算机之间使用共享介质，即总线传输数据帧，只能实现半双工通信。为确保多个站点对总线的有效竞争，CSMA/CD 协议采用了载波帧听、多路访问或冲突检测技术。

载波帧听：多个计算机在发送数据帧前，先帧听总线是否空闲，如果空闲，则发送数据帧；否则等待，再继续帧听，直到总线空闲。载波帧听实际上就是利用电子技术检测总线上有没有其他计算机发送的数据信号。

多路访问：多个计算机以多点接入方式连接在一根总线上，它们都有访问总线的权利。

冲突检测：计算机在发送数据帧的过程中，要不断检测线路是否有冲突，如果发生冲突，则停止发送数据帧，同时给发生冲突的线路发送一个阻塞信号，随机等待一段时间，继续帧听总线是否空闲。实际上，计算机是边发送数据帧边检测总线上的信号电压大小的。当多个计算机同时在总线上发送数据帧时，总线上的信号电压的摆动值将会增大（互相叠加）；当一个站点检测到总线上的信号电压的摆动值超过了一定门限值时，就认为总线上至少有两个站点在同时发送数据帧，这就表明产生了冲突。所谓冲突就是发生了碰撞，因此冲突检测也称碰撞检测。

发送数据帧的流程分为以下两个阶段。

第一阶段：当站点需要发送数据帧时，先帧听总线是否空闲

（1）如果总线空闲，则站点立即发送数据帧。

（2）如果总线忙碌，则继续帧听，直到总线变为空闲，再发送数据帧。

（3）在发送数据帧的过程中，要边发送边冲突检测。

第二阶段：如果在发送数据帧的过程中检测到了冲突，则应进行如下操作。

（1）立即停止发送该数据帧。

（2）给总线上发送一串阻塞信号，告诉其他站点总线发生冲突。

（3）等待一段随机时间（利用二进制指数退避算法），再重新争用总线，然后重复上面步骤，重发该数据帧。

冲突检测的基本方法是发送方在向总线发送数据帧时，先检测总线上的信号电压，如果信号电压的摆动值超过了一定门限值，则表明发生冲突。另一种方法是编码违例判别法，发送方在发送数据帧时，接收线路上的信号，并检测线路上的信号是否符合曼彻斯特编码。如果线路上的信号违背曼彻斯特编码，则说明总线发生冲突，应立即停止发送数据帧。

在共享式以太网中，因为信号在总线上传输时存在传播延迟的情况，所以发送方在判断总线空闲时会得到错误结果，造成总线上同时传输多路信号，从而发生信号冲突。冲突的结果是所有发送方发送的数据帧都变得无用了。

情况一：A、B 两个站点都检测到总线空闲，同时给总线上发送数据帧，结果两路信号在总线上发生冲突，发生冲突的位置在总线中点。假设 A 站点到 B 站点的信号传播延迟时间为 τ，则总线上发生冲突的时刻为 $\tau/2$，A、B 两个站点分别在 τ 时刻可检测到冲突发生，如图 3-8 所示。

图 3-8 情况一冲突发生示意图

情况二：A 站点先检测到总线空闲，向 B 站点发送一数据帧，当该数据帧还没有传播到 B 站点时（传播延迟），B 站点检测到总线空闲，向 A 站点发送数据帧，这时会发生冲突，发生冲突的位置在靠近 B 站点的一侧。假设 A 站点到 B 站点的传播延迟时间为 τ，而 B 站点在 $t = \tau - \delta$ 时刻有数据帧要发送，并检测到总线空闲，则总线发生信号冲突的时刻为 $\tau - \delta/2$，A、B 两个站点分别在 $2\tau - \delta$ 时刻和 τ 时刻可检测到冲突发生（δ 指此时 A 站点发送的数据帧传播到 B 站点所用的时间），如图 3-9 所示。

图 3-9 情况二冲突发生示意图

通过以上分析可知，使用 CSMA/CD 协议的共享式以太网不能进行全双工通信，而只能进行双向交替半双工通信。计算机发送数据帧后的一小段时间内，存在着遭遇冲突的可能性。由于网络冲突的不确定性，使整个以太网平均数据传输率远小于以太网最高数据传输率。

在上述两个例子中，最远、最先发送数据帧的 A 站点在发送数据帧时开始计时，最多经过 $2\tau - \delta$ 时间才可检测到发送的数据帧是否遭遇了冲突，如果 δ 取无穷小量，则可以得到如下结论。

结论 1：在共享式以太网中，任何两个计算机之间最多经过 2τ（两倍的端到端往返传播时延）时间才可检测到发送的数据帧是否遭遇了冲突（从发送数据帧时开始计时）。

结论 2：经过 2τ 时间还没有检测到冲突，才能肯定这次发送不会发生冲突，即该数据帧不会与之后发送的数据帧发生冲突。

因此，以太网的端到端往返传播时延 2τ 称为冲突窗口（争用期）。在标准以太网中，取 $2\tau=51.2\mu s$ 为冲突窗口的大小。对于标准以太网，在冲突窗口内可发送 512 比特，即 64 字节数据帧，由于一检测到冲突就立即停止发送，因此这时已经发送出去的数据一定大于 64 字节。如果发生冲突，那一定是在发送的前 64 字节数据帧之内。标准以太网在发送数据帧时，若前 64 字节数据没有发生冲突，则后续的数据也不会发生冲突。以太网必须规定最短有效帧长为 64 字节，如果小于 64 字节，则必须使用填充位使其等于 64 字节。凡帧长小于 64 字节的数据帧都是由于冲突而异常中止的无效帧（碎片帧）。

为了强化冲突现象，当发送数据帧的计算机检测到发生了冲突时，需要进行以下操作。

（1）立即停止发送该数据帧。

（2）发送 4 个字节的人为干扰信号，即阻塞信号，以便让总线上所有其他站点都知道现在发生了冲突。

（3）随机等待一段时间（利用二进制指数退避算法），再重新争用总线；一旦争用到总线，立即重发发生冲突的数据帧。

二进制指数退避算法的步骤如下。

第一步：确定基本退避时间，一般取冲突窗口时间 $2\tau = 51.2\mu s$。

第二步：K 为冲突次数，当 $K \leq 16$ 时，则 $j = Min（K, 10）$。

第三步：从整数集合 $[0,1,\cdots,(2^j-1)]$ 中随机抽取一个数，记为 r。

第四步：随机等待时延为 $r \times 2\tau$。

第五步：当冲突次数达到 16 次仍不能成功时，则丢弃该帧，并向高层发送报告。

注意：当冲突次数大于或等于 10 且小于或等于 16 时，r 的取值范围为 0 ~ 1023。

3.3 共享式以太网数据链路层协议工作效率分析

在共享式以太网中，一个发送站点可以连续发送多个数据帧，数据帧间最小间隔为 9.6μs，相当于 96 比特数据的发送时间，如图 3-10 所示。因此，一个发送站点在检测到总线空闲后，还要等待 9.6μs 才能再次发送下一个数据帧。这样做的目的有两个：一是使接收站点在接收到数据帧时，有一定的时间缓存数据帧，并及时处理，将数据帧的数据部分交付给上层协议，以便做好接收下一个数据帧的准备；二是通过违例码区分不同数据帧。

图 3-10　共享式以太网帧间最小间隔

在共享式以太网工作环境下，总线上连接的站点越多，同一时间有多个站点发送数据帧的概率就越大，总线发生冲突的可能性也增加，从而导致总线的利用率下降。所以总线利用率会随着总线上连接站点数量的增加而降低。在总线上连接站点数量确定的情况下，随着各站点每次发送数据帧长度的增加，总线发生冲突的可能性也增加，从而造成总线利用率下降，所以在发送站点数量确定的情况下，总线利用率会随着站点发送数据帧长度的增加而降低，如图 3-11 所示。

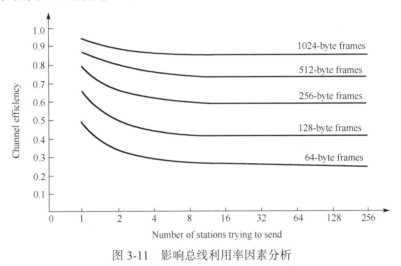

图 3-11　影响总线利用率因素分析

3.4　网络适配器 MAC 地址

严格来说，MAC 地址应当是网络适配器地址（或网络适配器标识符 EUI-48），用来代表每一个站点的名称或标识。MAC 地址共 48 比特位，IEEE 的注册管理机构 RA 负责向厂家分配地址字段的前 3 个字节，即高位的 24 个比特位。例如，华为标识为 0X00E0FC；思科标识为 0X00000C。MAC 地址中的后 3 个字节，即低位的 24 个比特位由厂家自行指派，只要确保唯一性即可。一个厂商的地址块最多有 2^{24} 个不同的 MAC 地址。网络适配器从网络上每接收一个数据帧，都要依次在数据链路层检查其是否为碎片帧。在接收的数据帧中，目的 MAC 地址是否与本站点地址匹配，CRC 检测是否正确，帧长是否是字节数的整数倍，以上检测只要有一项错误，则该帧丢弃，否则接收。发往本站的数据帧一般包括以下 3 种。

（1）单播（unicast）帧：用于一对一通信模式。

（2）组播（multicast）帧：用于一对多通信模式。

（3）广播（broadcast）帧：用于一对全体网段网络接口通信模式，目的 MAC 地址为 0XFFFFFFFFFFFF。

3.5 工业以太网数据帧发送和接收流程

3.5.1 工业以太网数据帧发送流程

在工业以太网中，站点在发送数据帧之前，先要检测总线是否空闲，以确定总线上是否有其他站点正在发送数据帧；如果总线空闲，则可以发送；如果总线忙碌，则要继续检测，一直检测到总线空闲方可发送。站点在发送数据帧时，还要持续检测总线是否发生冲突，一旦检测到冲突发生，立即停止发送数据帧，并向总线发出一串阻塞信号来加强冲突，以便让总线上其他站点都知道总线上发生冲突。这样，总线带宽不会因传送已损坏的数据帧而被白白浪费。总线上冲突发生后，站点应随机延迟一个时间，再去争用总线。通常采用的延迟算法是二进制指数退避算法。这个算法是按照后进先出的次序控制的，即未发生冲突或很少发生冲突的数据帧可以优先发送，而发生过多次冲突的数据帧，发送成功的概率反而小了。CSMA/CD 协议中发送站点发送数据帧流程如图 3-12 所示。

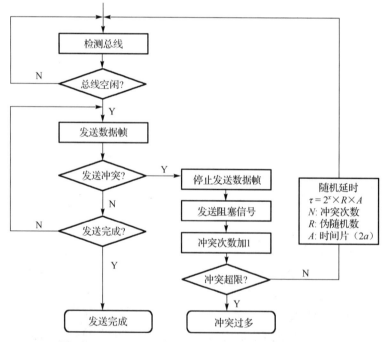

图 3-12 CSMA/CD 协议中发送站点发送数据帧流程

3.5.2 工业以太网数据帧接收流程

在工业以太网中，当一个站点发送数据帧时，其他站点先检测总线状态，当总线上有载波信号且变成活跃状态时，将启动数据帧接收过程，如图 3-13 所示。每个接收站点必须对接收的数据帧进行有效性检查，具体如下。

（1）滤除因冲突而产生的"帧碎片"，即当接收的数据帧的长度小于最小帧长限制（64

个字节）时，则认为该帧是不完整的帧而丢弃。

（2）检查数据帧的目的 MAC 地址字段是否与本站点地址匹配。地址匹配分两种情况：如果目的 MAC 地址为单播地址，则两个地址必须完全相同；如果目的 MAC 地址为组播地址或广播地址，则认为地址是匹配的，因为 MAC 子层没有能力处理组播地址或广播地址的数据帧，所以必须先接收下来，然后提交给上层协议来处理。如果地址不匹配，则说明数据帧不是发送给本站点的，应丢弃该帧。

（3）对数据帧进行 CRC 校验，如果 CRC 校验有错，则丢弃该帧。

（4）对数据帧进行长度检验，接收到的数据帧帧长必须是 8 比特位的整数倍，否则丢弃该帧。

（5）保留有效的数据帧，去除数据帧帧头和帧尾后，将数据帧的数据部分提交给 LLC 子层。

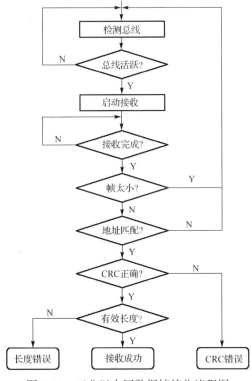

图 3-13　工业以太网数据帧接收流程图

3.5.3　MAC 子层与相邻层的接口

MAC 子层定义了两个与相邻层的接口，具体如下。

（1）MAC 子层与 LLC 子层之间的接口。MAC 子层通过该接口向 LLC 子层提供 LLC 帧的发送与接收服务。该接口定义了两个功能，即帧发送和帧接收功能。LLC 子层可以通过该接口使用 MAC 子层设施来发送和接收 LLC 帧。

（2）MAC 子层与 PLS 层之间的接口。PLS 层通过该接口向 MAC 子层提供数据帧的发送与接收服务。该接口定义了两个功能,即位发送与位接收功能,以及 3 个状态变量:冲突检测、载波监听和发送正在进行中。MAC 子层通过该接口使用物理层设施,并根据物理层提供的总线状态,对总线访问实施相应的控制。

10BASE5 参数值如表 3-1 所示。其中,最大重传次数表示当发生 16 次冲突后,MAC 子层便停止动作,并向高层报告错误;退避极限表示当发生 10 次冲突后,随机等待的最大时隙被固定在 1023,而当冲突次数小于 10 时,等待时隙则从 2^i-1 中随机选出。

表 3-1 10BASE5 参数值

参　数	数值
时间片大小（Slot Time）	512 比特位时间（相当于 51.2μs）
帧间间隔（Inter Frame Gap）	9.6μs
最大重传次数（Attempt Limit）	16
退避极限（Back Off Limit）	10
阻塞信号大小（Jam Size）	32 比特位
最大帧长（Max Frame Size）	1518 字节
最小帧长（Min Frame Size）	512 比特位（64 字节）
地址字段长度（Address Size）	48 比特位

以太网标准中的帧格式与 IEEE 802.3 标准中的帧格式基本相同,只是 IEEE 802.3 标准中的帧格式中的帧长度（FL）字段在以太网标准的帧格式中被定义为帧格式（FT）字段。在其他方面,IEEE 802.3 标准的 CSMA/CD 协议非常接近于以太网标准。事实上,两者之间的大多数差异已经在 CSMA/CD 协议的高版本中得到解决。各种以太网 MAC 子层协议都采用 CSMA/CD 协议,而在物理层协议上则各有不同。

3.6　共享式以太网数据帧发送源程序

共享式以太网数据帧发送源程序如下。

```
#include "stdafx.h"
#include "ConsoleApplication3.h"

#ifdef _DEBUG
#define new DEBUG_NEW
#endif

CWinThread *thread1, *thread2;
DWORD ID1, ID2, Bus = 0;
UINT aThread(LPVOID pParam);
UINT bThread(LPVOID pParam);
using namespace std;
int _tmain(int argc, TCHAR* argv[], TCHAR* envp[])
{
    int nRetCode = 0;
```

```cpp
    if (!AfxWinInit(::GetModuleHandle(NULL), NULL, ::GetCommandLine(), 0))
    {
        cerr << _T("Error:MFC initialization failed") << endl;
        nRetCode = 1;
    }
    else
    {
        //启动线程 aThread，获取该线程 ID
        thread1 = AfxBeginThread(aThread, NULL);
        ID1 = thread1->m_nThreadID;
        //启动线程 bThread，获取该线程 ID
        thread2 = AfxBeginThread(bThread, NULL);
        ID2 = thread2->m_nThreadID;
        getchar();
    }
    return nRetCode;
}

UINT aThread(LPVOID pParam)
{
    int i = 0;
    //设置最大冲突次数
    int CollisionCounter = 16;
    double collisionWindow = 0.005;
    int randNum = rand() % 3;
    //模拟不同帧长对总线利用率的影响
    Int sendFrametime = FrameLength*0.1;
Loop:
    if (Bus == 0)    //总线空闲
    {
        Bus = Bus | ID1;
        //模拟数据帧发送时间
        Sleep(sendFrametime);
        if (Bus == ID1)
        {
            printf("%d Send Success\n", ID1);
            Bus = 0;
            CollisionCounter = 16;
            Sleep(rand() % 10);
            i++;
            printf("主机 a 发送成功次数= %d\n", i);
            if (i<5)
            {
                goto Loop;
            }
        }
```

```c
        else
        {
            printf("%d Send Collision\n", ID1);
            Bus = 0;
            CollisionCounter--;
            if (CollisionCounter>0)
            {
                Sleep(randNum*(int)pow(2, (CollisionCounter>10) ?
                    10 : CollisionCounter)*collisionWindow);
                goto Loop;
            }
            else
            {
                printf("%ld Send Failure\n", ID1);
            }
        }
    }
    else
    {
        goto Loop;
        return 0;
    }
}

UINT bThread(LPVOID pParam)
{
    int i = 0;
    int CollisionCounter = 16;
    double collisionWindow = 0.005;
    int randNum = rand() % 3;
    //模拟不同帧长对总线利用率的影响
    Int sendFrametime = FrameLength*0.1;
Loop:
    if (Bus == 0)
    {
        Sleep(2);
        Bus = Bus | ID2;
        Sleep(sendFrametime);
        if (Bus == ID2)
        {
            printf("%d Send Success\n", ID2);
            Bus = 0;
            CollisionCounter = 16;
            Sleep(rand() % 10);
            i++;
            printf("主机a发送成功次数= %d\n", i);
```

```
            if (i<5)
            {
                goto Loop;
            }
        }
        else
        {
            printf("%d Send Collision\n", ID2);
            Bus = 0;
            CollisionCounter--;
            if (CollisionCounter>0)
            {
                Sleep(randNum*(int)pow(2, (CollisionCounter>10) ?
                    10 : CollisionCounter)*collisionWindow);
                goto Loop;
            }
            else
            {
                printf("%ld Send Failure\n", ID2);
            }
        }
    }
    else
    {
        goto Loop;
        return 0;
    }
}
```

思考题：

1. 利用以上程序，模拟仿真站点发送数据帧的长度对总线利用率的影响，以获得仿真数据，并通过画图的方法进行数据分析，找出原因。

2. 利用以上程序，模拟总线上连接站点数量的大小对总线利用率的影响，以获得仿真数据，并通过画图的方式进行数据分析，找出原因。

第 4 章 网络通信协议的设计与实践

4.1 引 言

从传统意义上看,数据链路层的基本功能是采用流量控制和差错控制技术将不可靠的物理链路变成可靠的数据链路。因此,不论是 HDLC 规程还是 LLC 子层协议,都采用了某种流量控制和差错控制技术来实现数据链路的同步控制和可靠传输。

数据链路层流量控制的研究对象是接收方接收缓存队列,解决的问题是如何利用缓存队列来保证其向上层交付速率大于或等于接收速率,目的是确保接收缓存不溢出。由于接收速率等于发送方发送速率,因此,流量控制技术实质上是一种协调发送方的发送速率和接收方的接收速率的一致性的数据传输同步技术,一般是利用接收缓存,然后用接收速率控制发送速率。在数据传输过程中,发送方将数据封装成数据帧发送出去,发送速率是指生成和发送数据帧的速率,是以每秒发送的帧数(f/s)或字节数为速率单位的。接收方先将接收的数据帧暂存在接收缓存区中,再进行必要的处理,如帧头有关字段处理等。向上层交付速率是指从接收缓存区取出数据帧进行处理并成功交付给上层的速率,它是以每秒处理的帧数为速率单位的。如果接收速率大于交付速率,则接收方会来不及处理接收的数据帧,从而产生接收缓存区的数据溢出,造成数据帧丢失,这种现象称为同步失调。如果接收速率远小于交付速率,则接收方会一直处于等待状态,造成介质空闲,使介质利用率过低。通过流量控制技术可以有效地解决同步失调和高效利用介质的问题。

4.2 网络通信协议可靠性原理

4.2.1 检错与纠错机制

通信系统的基本任务是高效而无差错地传送数据,但在任何一种通信线路上都不可避免地会存在一定程度的噪声,因而产生比特差错。在信道中存在两种比特差错:随机差错和突发差错,其产生原因和特点如表 4-1 所示。实际上,网络中还存在另一种由传输引起的差错,如乱序、重复、丢失等,这类差错称为传输差错。由于数据链路层研究的是两个相邻网络节点之间的通信可靠性问题,因此不存在网络交换设备的转发或路由,所以数据链路层仅存在比特差错,不存在传输差错。

第 4 章 网络通信协议的设计与实践

表 4-1 不同类型的比特差错比较表

种类	随机差错	突发差错
来源	随机热噪声（白噪声）	冲击噪声
特点	信道固有的、持续存在的 码元的差错是独立的，和前后的码元无关	外界的因素，持续时间短、突发性的 差错具有相关性，数据传输中产生差错的主要原因

突发差错长度是指差错发生的第一个码元到最后一个码元间的所有码元的数。误码率是衡量物理信道的通信质量的一个指标，计算公式如下。

$$P_\varepsilon = \frac{错误比特数}{总比特数}$$

为了在数据链路层实施检错，一般采用差错控制编码：数据（k 比特位）+校验码（r），其中，校验码根据功能又分为检错码和纠错码。检错码是指可以自动以一定概率发现差错的编码，如奇偶校验码和 CRC 校验码。纠错码是指既能发现差错又能自动纠正比特差错的编码，如海明威编码等。编码效率（R）是指编码中有效信息位所占的比例：$R=k/(k+r)$。漏检率是指某比特位出错但接收者无法检测到的概率。

奇偶校验码是通过增加一比特位校验位使得数据+校验码中"1"的个数为奇数或偶数来实施检错的。奇校验码是指增加一比特位校验位（可为 0 或 1）使得数据+校验码中"1"的个数为奇数的编码。偶校验码是指增加一比特位校验位（可为 0 或 1）使得数据+校验码中"1"的个数为偶数的编码。如果数据串行传输，则奇偶校验码可以智能检错，但无法实施纠错功能，当接收方发现错误后只能请求重发。虽然奇偶校验码只能检测出信息传输过程中的部分误码（奇数位误码能检出，但偶数位误码不能检出），但由于通过奇偶校验码来实施检错比较简单，因此得到了广泛使用。如果基于两路奇偶校验码，采用数据并行传输线路，则可以部分实施纠错功能，如图 4-1 所示。

	字符1	字符2	字符3	字符4	字符5	字符6	字符7	字符8	校验字符
b1	1	1	1	1	1	1	1	1	1
b2	0	1	1	0	0	0	1	1	0
b3	0	0	1	0	0	0	0	0	1
b4	0	0	0	0	1	1	0	0	0
b5	0	0	0	0	0	0	0	0	0
b6	0	0	1	0	0	1	0	0	0
b7	1	1	1	1	1	1	1	1	1
校验	0	1	1	1	0	0	1	1	0

图 4-1 基于两路奇偶校验码的数据并行传输

另一种检错码为冗余校验码，也称为 CRC 校验码。在 CRC 校验码的计算中，假设校验码为 n 比特位，发送方首先选择要发送的数据为被除数，该被除数需要左移 n 位或在数据右边补 n 个 0；然后选择一个预定的二进制码（$n+1$ 位）作为除数；再利用二进制模二除法运算，计算得到一个余数（n 比特位），该余数即可作为 CRC 校验码随数据帧发送给接收方，如图 4-2 所示。接收方以发送方发送的数据和 CRC 校验码为被除数，选同样的预定的二进制码为除数，如果余数为 0，表明该数据帧在传输过程中接近概率 1 没有发生差错，否则表明发生差错。

图 4-2 基于数据检错的编码格式

预定的二进制码（除数）一般是一个标准的国际编码，用户不能任意规定，其特点是最高位和最低位为 1；一般以多项式形式表示，称为生成多项式 G(x)；局域网（以太网、令牌环网）采用 CRC-32，HDLC 规程采用 CRC-CCIT，ATM 协议采用 CRC-8、CRC-10 和 CRC-12；发送数据的长度大于生成多项式表示的二进制码的长度。

CRC-8：$G(x)=x^8 + x^2 + x + 1$。
CRC-10：$G(x)= x^{10} +x^9 + x^5 +x^4 + x + 1$。
CRC-12：$G(x)=x12 +x^{11} +x^3 +x^2 + x + 1$。
CRC-16：$G(x)=x^{16} +x^{15} +x^2 + 1$。
CRC-CCIT：$G(x)=x^{16} +x^{12} + x^5 + 1$。
CRC-32：$G(x)=x^{32}+x^{26}+x^{23}+x^{22}+x^{16}+x^{12}+x^{11}+ x^{10} +x^8 +x^7+x^5+x^4 + x^2+x+1$。

CRC 校验码可以检测出全部单一比特位出现的差错；只要 G(x)中含有一个至少 3 项的因子，就可以检测出所有两个比特位出现差错的情况；只要 G(x)中含有因子(x+1)，就可以检测出全部奇数位出现差错的情况；r 为生成多项式的最高幂次，CRC 校验码可以检测出差错长度小于或等于 r 的所有突发性差错，并以 $1-(1/2)^{r-1}$ 的概率检测出差错长度大于 r 位的突发性差错。如果 $r=16$，那么该 CRC 校验码能检测出全部差错长度小于或等于 16 位的突发性差错，并以 $1-(1/2)^{16-1}=99.997\%$ 的概率检查出差错长度大于 16 位的突发性差错，漏检概率为 0.003%。共享式以太网在计算 CRC 校验码时，生成多项式采用 CRC-32，协议计算源代码如下。

```
u_int32_t crc32_table[256];
//generate table
void generate_crc32_table()
{
    int i, j;
    u_int32_t crc;
    for (i = 0; i < 256; i++)
    {
        crc = i;
        for (j = 0; j < 8; j++)
        {
            if (crc & 1)
                crc = (crc >> 1) ^ 0xEDB88320;
            else
                crc >>= 1;
        }
        crc32_table[i] = crc;
    }
}
```

```
u_int32_t calculate_crc(u_int8_t *buffer, int len)
{
    int i, j;
    u_int32_t crc;
    crc = 0xffffffff;
    for (i = 0; i < len; i++)
    {
        crc = (crc >> 8) ^ crc32_table[(crc & 0xFF) ^ buffer[i]];
    }
    crc ^= 0xffffffff;
    return crc;
}
```

4.2.2 流量控制机制

完全理想化的数据传输信道基于以下两个假设。

假设 1：数据帧在理想信道上传输时既不会发生差错（比特差错），也不会丢失（传输差错），这涉及差错控制问题。

假设 2：不管发送方以多快速率发送数据帧，接收方总是来得及接收的，并可及时将数据帧的数据部分交付给上层协议，这涉及发送速率和交付速率匹配问题，主要通过流量控制来实现。

根据假设 2，接收方向上层交付数据帧的速率永远不会低于发送方发送数据帧的速率。在实际网络环境中，以上两个假设均不成立。如果数据帧在相邻两个网络节点间的传输过程中出现了差错，那纠正差错的方法有两种。第一种方法是自动重发请求（Automatic Request for Repeat，ARQ）协议。在该方法中，接收方检测错误并丢弃出错帧，发送方重传出错帧，但前提是发送方知道哪个数据帧出错了，所以需要给每个数据帧编号，并缓存已发送的数据帧。第二种方法是前向纠错法（Forward Error Correction，FEC）。在该方法中，接收方利用纠错码（海明威码）不仅可以检测差错，而且可以知道差错的位置，从而纠正差错。前向纠错法的优点是无须重发出错帧；缺点是编码效率低，算法比较复杂，实现比较困难，因此很少使用。

网络通信时会出现比特差错和传输差错，目前数据帧的可靠传输方法主要采用序号+确认反馈+超时重传机制的方式。为了解决发送方发送速率和接收方交付速率的匹配问题，一般利用接收方交付速率来控制发送方的发送速率，这涉及流量控制问题；在数据链路层一般采用固定大小的滑动窗口技术来实现流量控制。现在的 ARQ 协议除支持差错控制之外，同时可以实现流量控制，具体可分以下为两类。

（1）停止-等待 ARQ 协议（以下简称停止-等待协议）。

（2）连续 ARQ 协议：后退 N 帧协议和选择重发协议。

传统的停止-等待协议仅具有流量控制功能，如图 4-3 所示。发送方每发送一个数据帧，通过启动一个重发定时器，可以扩展为带差错控制的停止-等待协议。如果重发定时器超时，且发送方发送的数据帧对应的应答帧还没有收到，则表明该数据帧出现了比特差错，需要重发，否则发送下一个数据帧，如图 4-4（a）所示。在图 4-4（b）中，数据帧 DATA0 出现比特差错，接收方丢弃该数据帧并不发送应答帧，造成该数据帧对应的发送方的重发定时器超时而重发；在图 4-4（c）和 4-4（d）中，接收方会接收到重复的数据帧，但接收方根据数据帧序号，将后来重复的数据帧丢弃即可解决该问题。

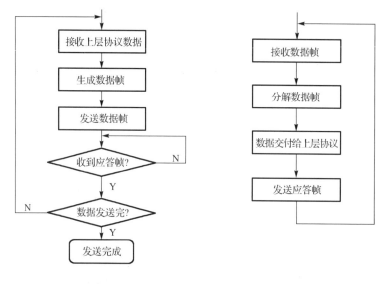

(a) 发送方工作流程　　　　　　　　(b) 接收方工作流程

图 4-3　传统的停止-等待协议工作流程

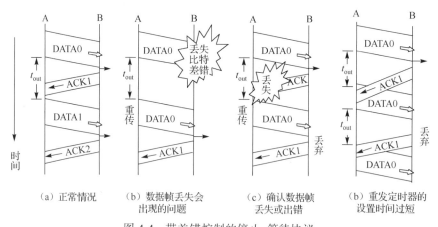

(a) 正常情况　　(b) 数据帧丢失会　　(c) 确认数据帧　　(b) 重发定时器的
　　　　　　　　　出现的问题　　　　丢失或出错　　　　设置时间过短

图 4-4　带差错控制的停止-等待协议

接收方接收到重复数据帧的主要原因是，发送方的重发定时器超时而产生数据帧重发。引起重发定时超时的原因如下。

（1）重复定时器设定时间过短，数据帧在传输过程中出现了丢失、比特差错或传输延迟时间过长。

（2）ACK 应答在传输过程中出现了丢失、差错或传输延迟时间过长。

在停止-等待协议中，通过引入 1 比特位的序号字段来解决数据帧的编号问题。数据帧中的发送序号以 0 和 1 交替的方式出现在数据帧中。一般情况下，发送方的重发定时器的时间设置策略为略大于从发完数据帧到确认收到数据帧所需的平均时间（Round Trip Time，RTT）。目前，如果数据链路层数据帧出现比特差错，则接收方的处理方式有以下两种。

方法一：采用鸵鸟策略，即不处理，发送方利用重发定时器超时重传来解决。

方法二：接收方成功接收到上一个数据帧后，启动一个接收定时器，如果接收定时器超时或数据出现差错，则发送一个 ACK 应答，要求发送丢失数据帧或差错数据帧。这种解决办法要占用接收方的大量资源，所以很少使用。

在停止-等待协议中，如果接收方连续接收到相同序号的数据帧，则表明发送方进行了超时重传。发送方在发送完数据帧时，必须在其发送缓存中暂时保留已发送的数据帧副本，这样才有条件在传输后进行重传。发送方只有接收到 ACK 应答，才可以清除数据帧副本。停止-等待协议的优点是实施简单，缺点是实施效率低，物理链路的利用率比较低，信道远远没有被数据比特填满。例如，一条线路带宽为 1.5MB/s，RTT=45ms，如果采用停止-等待协议，则发送方只能在一个 RTT 时间内发送一个数据帧；假设数据帧大小为 1 千字节，则发送方实际的发送速率为 182MB/s，相当于物理链路带宽的 1/8。造成物理链路带宽浪费的主要原因是，当发送方发送一个数据帧后，不能继续发送，需等待相应 ACK 应答；网络存在各种延迟（差错、无丢失或速率匹配）。物理链路带宽浪费的改进方法是，首先要提高物理链路利用率，发送方在等待接收方返回第一个 ACK 应答前，再连续发送 7 个数据帧，这也是连续 ARQ 协议的思想。

连续 ARQ 协议是指发送方可一次连续发送多个数据帧，在发送数据帧的同时等待 ACK 应答。该协议引入了滑动窗口技术，也称为滑动窗口协议。连续 ARQ 协议分为后退 N 帧协议和选择重发协议。后退 N 帧协议的工作原理如下。

（1）发送方在发送一个数据帧后，不是停下来等待 ACK 应答，而是可以连续发送多个数据帧，同时启动重发定时器。

（2）如果发送方发送的数据帧的重发定时器在超时前收到了接收方发来的 ACK 应答，则发送方可以通过滑动窗口技术继续发送后面的数据帧。

（3）如果发送方发送的数据帧的重发定时器已经超时，但还没有收到接收方发来的 ACK 应答，则从该帧开始的后续帧全部重发。

在图 4-5 中，发送方成功发送了 DATA0、DATA1 数据帧，但 DATA2 数据帧在传输过程中发生比特差错，所以即使 DATA3、DATA4、DATA5 数据帧没有差错，接收方仍会按照乱序错误而丢弃这些数据帧；因为接收方窗口大小为 1，接收方按照序号接收数据帧，所以只有数据帧序号与接收窗口序号一致时，才会接收数据帧，否则丢弃，原因是接收方没有空间缓存后面正确的数据帧。在图 4-5 中，由于 DATA2 数据帧发生了差错，接收方无法接收，也不可能向发送方发送 ACK 应答，因此发送方 DATA2 数据帧对应的重发定时器超时，发送方重发 DATA2、DATA3、DATA4、DATA5 数据帧；由于发送方发送窗口大于 1，所以可以连续发送多个数据帧，而接收方窗口等于 1，因此一次只能按照序号接收一个数据帧。在后退 N 帧协议中，如果发送方连续收到相同序号的 ACK 应答，则说明接收方在接收数据帧时一定出现了乱序。例如，接收方接收的数据帧出现丢失或差错，同时收到后续数据帧的情况。从技术角度，可以通过滑动窗口技术来解释后退 N 帧协议。在发送方和接收方的缓存区分别设定一个发送窗口和接收窗口。针对发送窗口，发送方只能发送窗口内的数据帧，并对发送方进行流量控制，窗口大小表示在没有收到接收方确认的情况下，发送方最多可以发送的数据帧个数，发送方每接收一个 ACK 应答，窗口向前滑动一个单位。针对接收窗口，接收窗口用来控制接收方可以接收的数据帧的个数，只有数据帧序号与接收窗口序号一致时才可以接收，否则丢弃；接收方正确接收并处理数据帧后，交付给上层，窗口向前滑动一个单位，并发送一个 ACK 应答给发送方。接收方对数据帧的处理流程如下。

（1）完整地接收数据帧，比较数据帧序号是否与接收窗口序号一致。

（2）对数据帧帧头和帧尾进行处理，看是否有比特差错。

（3）去除数据帧帧头和帧尾，将数据帧的数据部分按照协议类型交付给上层协议并继续处理。

（4）向发送方发送一个 ACK 应答。

（5）接收窗口向前滑动一个位置，准备接收下一个数据帧。

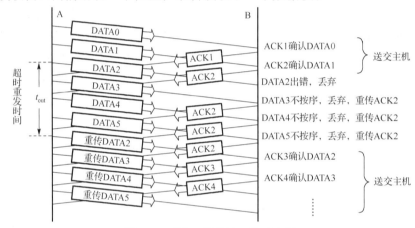

图 4-5　后退 N 帧协议工作示意图

在具体实践中，发送缓存和接收缓存可以采用循环队列的方式；窗口可以利用指针来指示不同位置，接收方一般采用累计应答，而不是一次一应答的方式。接收方对乱序帧的处理如图 4-6 所示。

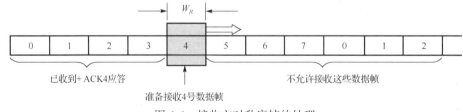

图 4-6　接收方对乱序帧的处理

在图 4-6 中，接收方准备接收 4 号数据帧，但发送方可能连续发送了 4 号、5 号、6 号、7 号数据帧。由于 4 号数据帧发生比特错误，因此接收方采用鸵鸟策略丢弃该数据帧；5 号、6 号、7 号数据帧的接收方即使正确接收了这些数据帧，也会将这 3 个数据帧当作乱序帧丢弃，并向发送方发送 3 个连续的 ACK4 应答。

在后退 N 帧协议中，如果序号字段为 k 比特位，则数据帧序号空间为 $[0, 2^k-1]$，发送窗口序号的最大值为 2^k-1，可保证该协议在任何情况下不会出现逻辑上的错误。例如，序号字段为 3 比特位，序号空间为 0、1、…7；发送窗口最大值为 7，即一次最多可连续发送 7 个数据帧。由此可见，发送方按照发送窗口的大小一次可连续发送多个数据帧，发送方每发送一个数据帧，都要设置该数据帧的重发定时器；如果重发定时器未超时且收到 ACK 应答，则立即将重发定时器清零，接收窗口向前滑动一个单位；如果重发定时器超时，也未收到 ACK 应答，则重发该数据帧后的所有数据帧。接收方按序号接收数据帧，当接收方接收到一个数据帧时，首先检测其正确性，然后排队处理，并将数据帧的数据部分交付给上层协议，这时接收窗口向前滑动一个单位，并发送一个 ACK 应答。后退 N 帧协议一般采用累计确认的方式，ACKn 表示确认 n–1 号前后所有数据帧均已成功接收，并期望下次接收 n 号数据帧。后退 N 帧协议支持捎带确认，对于全双工通信，接收方在给发送方发送数据时捎带确认，以提高通信效率。后退 N 帧协议中存在的问题是，将发生差错的数据帧后所有的数据帧重发，会造成物理链路带宽的浪费，产生该问题的主要原因是接收窗口等于 1，接收方无法缓存后面正确接收的数据帧，因此有过多重复的数据帧在网络上传输；改进方法：一方面，增大接收窗口，用于缓存已正确接收的数据帧，另一方面，发送方只重传出现差错的数据帧。由此产生了选择重发协议。

在选择重发协议中，通信双方约定：当发送方重发定时器超时时，仅重发发生差错的数据帧，而不重发自发生差错的数据帧起的后续各个序号的数据帧；接收方在接收到一个发生差错的数据帧时，仅丢弃发生差错的数据帧，其后的数据帧按照接收窗口序号依次缓存即可。这样约定的主要原因是接收方的接收窗口大于 1，接收方有缓存空间来暂存后面接收到的各个正确的数据帧，等接收方的接收窗口内所有的数据帧都确认正确无误后，接收方再将数据帧的数据部分交付给上层协议做进一步处理。

选择重发协议的物理链路有效利用率比较高，但需要较大的缓存空间，如图 4-7 所示。

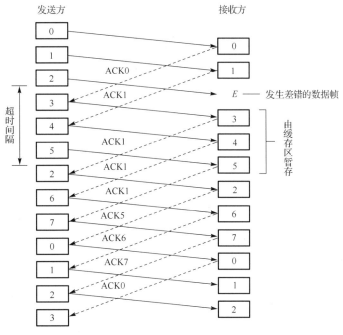

图 4-7 选择重发协议工作示意图

4.3 数据链路层通信协议设计

4.3.1 数据链路层通信协议设计要求

在数据链路层设计一个相邻两个网络节点之间的通信协议，具体要求如下。

（1）数据链路层通信协议既可以支持数据链路层通信的可靠性，又可以支持数据链路层通信的不可靠性，具体采用哪种通信模式由用户自己选择。

（2）可靠性要求：提供差错控制机制，检错采用 CRC-32，纠错采用序号+确认反馈+超时重传机制；支持流量控制机制，采用选择重发协议，其中序号字段为 3 个比特位，提供发送缓存区和接收缓存区并进行管理，提供发送窗口和接收窗口并进行管理。

（3）不可靠性要求：数据链路层通信协议支持不可靠通信服务，主要用于高质量物理链路。

（4）协议设计要求：从语法、语义和同步机制 3 个方面进行设计。

4.3.2 数据链路层通信协议语法设计

根据数据链路层通信协议设计要求，其语法结构如图 4-8 所示。该通信协议主要包括目的 MAC 地址字段、源 MAC 地址字段、控制字段、协议类型字段、数据字段、CRC-32 字段；其中，帧定界符不属于数据帧部分，其主要由物理层硬件对其进行处理。

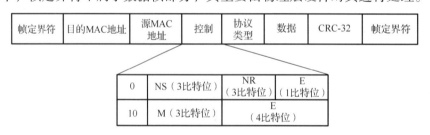

图 4-8 数据链路层通信协议语法结构

起始帧定界符与终止帧定界符各占一个字节，分别标识数据帧的开始和结束。相邻的两个数据帧，它们中间只需要一个帧定界符，帧定界符格式为"01111110"二进制序列。

目的 MAC 地址：48 比特位，标识发送方网络接口 MAC 地址。

源 MAC 地址：48 比特位，标识接收方网络接口 MAC 地址。

控制：定义数据帧的类型，实现差错控制和流量控制等功能。

协议类型：占一个字节，标识数据帧的数据部分为上层何种协议产生的协议数据单元，接收方依据该字段，将数据帧的数据部分交付给相应的上层协议进行下一步处理。

数据：发送方填写的上层协议数据单元，46～1500 字节。

4.3.3 数据链路层通信协议语义设计

为了支持可靠通信和不可靠通信两种通信模式，利用控制字段，将数据链路层数据帧划分为 2 种类型：信息 I 帧、监控 S 帧，如图 4-8 所示。

信息 I 帧：控制字段第一个比特位为 0，标识可靠通信中信息 I 帧，NS 为发送序号，NR 为应答序号；E 为扩展位（0/1）。

监控 S 帧：控制字段前两个比特位分别为 1、0；M 占 3 比特，用于标识不同控制帧类别；E 为扩展位。

M = 000：将通信模式设置为可靠通信，采用带差错控制的停止-等待协议。

M = 001：将通信模式设置为可靠通信，采用后退 N 帧协议。

M = 010：将通信模式设置为可靠通信，采用选择重发协议。

M = 011：将通信模式设置为不可靠通信。

M = 100：在可靠通信模式下，标识有序号的 ACK 应答，扩展位 E 为应答序号。

M = 101：无编号 UA 应答，在设置通信模式时使用。

M = 110：将数据链路状态设置为初始状态。

M = 111：通信结束，释放数据链路资源。

4.3.4　数据链路层通信协议同步机制设计

数据链路层通信协议同步机制是指事件发生的顺序，其同步关系可分为可靠通信和不可靠通信两种，可以采用时序图、流程图、伪代码及状态机等多种方式来描述。用时序图描述数据链路层通信协议同步关系如图 4-9 所示。

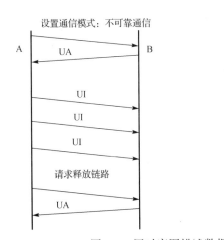
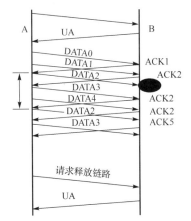

图 4-9　用时序图描述数据链路层通信协议同步关系

1. 不可靠通信模式下发送方发送数据帧伪代码

```
//建立数据链路，设置通信模式为不可靠通信模式
（1）向接收方发送"设置不可靠通信模式请求"
（2）接收到对方的 UA 应答
//通信前准备：资源初始化
（3）发送方初始化发送缓存队列
//通信阶段：不可靠通信模式
（4）从上层接收数据（可以模拟从文件读取数据），长度在 46～1500 字节
```

（5）进程 1 将数据封装成无编码信息帧 UI，并把数据帧缓存在发送缓存队列

（6）进程 2 从发送缓存队列中每隔 RTT 时间取出一个完整数据帧并发送，然后从发送缓存队列中删除该数据帧

（7）直到进程 2 发送完发送缓存队列中所有数据帧

//释放数据链路

（9）释放发送缓存队列，发送释放链路请求

（10）接收到对方 UA 应答，说明通信结束

2. 不可靠通信模式下接收方接收数据帧伪代码

//建立数据链路

（1）接收到"发送请求"，设置通信模式为不可靠通信模式

（2）发送 UA 应答给发送方

//通信前准备：初始化

（3）接收方初始化接收缓存队列

//通信阶段：不可靠通信方式

（4）进程 1 从下层接收到数据帧，检查目的 MAC 地址是否正确，若正确则存储在接收缓存队列，否则丢弃

（5）进程 2 从接收缓存队列中取出数据帧，并进行数据长度检查（46～1500 字节），然后通过校验码检查，如果均正确，则去掉帧头和帧尾，然后把数据帧的数据部分按照协议类型交付给上层协议（或写到接收文件中），否则丢弃该数据帧

（6）进程 1 继续从下层接（物理层）收数据帧，并存储在接收缓存队列，直到数据帧接收完毕

（7）进程 2 处理接收缓存队列中的所有数据帧，并将数据帧的数据部分交付给上层协议，直到接收缓存队列中的数据帧处理完毕

//释放数据链路

（8）接收对方发送的释放链路请求

（9）释放接收缓存队列等系统资源

（10）发送 UA 应答给发送方，说明通信结束

3. 可靠通信模式下发送方发送数据帧伪代码 （流量控制采用选择重发协议）

//建立数据链路，设置通信模式为可靠通信模式

（1）向接收方发送"设置可靠通信模式请求"，用户采用带差错控制的停止-等待协议、后退 N 帧协议或选择重发协议等

（2）接收到对方的 UA 应答

//通信前准备：资源初始化

（3）发送方初始化发送缓存队列和发送窗口，并进行管理

//通信阶段：如果采用可靠通信方式

（4）进程 1 从上层接收数据（从文件读取数据），长度为 6～1500 字节

（5）进程 1 将数据封装成信息 I 帧，并把数据帧缓存在发送缓存队列

（6）初始化发送窗口大小为 4

（7）进程 2 从发送窗口中每隔 RTT/4 时间发送一个数据帧，并启动重发定时器

（8）如果发送窗口中数据帧的重发定时器未超时且收到 ACK 应答，则从发送缓存队列删除该数据帧，窗口向前滑动一个单位，继续发送窗口内新的数据帧

（9）如果发送窗口内数据帧的重发定时器超时且未收到ACK应答，则仅重发该数据帧

（10）如果数据帧重发次数等于7次，则通信结束，GO TO(12)

（11）如果发送缓存队列中的数据帧未发送完，GO TO(7)

//通信结束前，释放数据链路

（12）释放发送缓存队列和发送窗口

（13）发送释放链路请求

（14）收到对方UA应答，说明通信结束

4．可靠通信模式下接收方接收数据帧伪代码（选择重发协议）

//建立数据链路，并设置通信模式

（1）接收到"发送请求"，设置通信方式为可靠通信模式（流量控制根据请求报文内容确定）

（2）发送UA应答给发送方

//通信前准备：初始化

（3）接收方初始化接收缓存队列和接收窗口，并进行管理

//通信阶段：如果采用可靠通信方式

（4）进程1从下层接收数据帧，如果数据帧的序号在接收窗口的序号内，并且目的MAC地址正确，则进行长度检查（46～1500字节）、校验检查，如果均正确，则将该数据帧缓存在接收缓存队列中

（5）如果进程1接收的数据帧在接收窗口内，目的MAC地址、长度、校验检查均正确，但数据帧的序号乱序，则该数据帧缓存在接收缓存队列（接收窗口内），发送重复ACK应答；否则丢弃

（6）进程2从接收窗口取出一个数据帧，去掉帧头和帧尾，把数据部分交付给上层（写到接收文件中）协议进行下一步处理，接收窗口向前滑动一个单位，并发送ACK应答给发送方

（7）直到数据帧接收完毕

//释放数据链路

（8）接收对方发送的释放链路请求

（9）释放接收缓存队列及接收窗口等系统资源

（10）发送UA应答给发送方，说明通信结束

4.4 数据链路层可靠通信协议实现

4.4.1 编程接口Winpcap

Winpcap是一个基于Win32平台的开源库，通常用于捕获网络上的数据包并进行统计分析，它可用于Windows系统下的直接网络编程，如捕获原始数据包，即没有经过操作系统的网络协议处理过的数据包。Winpcap提供了以下功能。

（1）捕捉原始数据包，一般是数据帧。

（2）在数据帧交付给上层协议前，根据用户指定的规则过滤数据帧。

（3）将原始数据帧通过网络接口发送出去。

（4）收集并统计网络流量信息。

基于Winpcap开发的网络应用程序，可用于网络及协议分析、网络监控、网络入侵检测系统（NIDS）、网络扫描及其他安全工具等方面。Winpcap的内部结构包括一个内核级

别的 Packet filter、一个底层的 DLL（Packet.dll）和一个高级的独立于系统的 DLL（Wpcap.dll）。Winpcap 编程中涉及的几个重要函数如下。

1）pcap_findalldevs_ex()

作用：获取所有能被 pcap_open() 打开的网络设备（网络接口）。

参数：pcap_if ** alldevsp 是自定义类型指针变量，用以表示网络接口列表；char *errbuf 是字符型指针变量，是承载错误信息的字符串。

返回值：在 int 型中，–1 代表函数打开失败；0 代表函数打开成功。

2）pcap_t* pcap_open ()

作用：打开或启动一个网络接口以捕获数据。

参数：const char * source 是字符型指针变量，包含要打开的网络接口名称，并且不能为 NULL。int snaplen 是整型变量，存储的值为需要保留的数据包的长度。对每一个过滤器捕获到的数据包，"snaplen"变量的首字节的内容将被存储至缓存区。例如，如果"snaplen"变量值为 10，那么仅每个数据包的首个 10 字节的内容被缓存。简单来说，就是从每一个数据包的开头到"snaplen"变量值的那段内容将被缓存。int flags 是保存一些由于抓包需要的标志。int read_timeout 是整型变量，单位为毫秒，其作用是在接收一个数据包时，读取操作不需立即返回，而是等待一段时间，然后一次性读取该时间段内到来的所有数据包；"read_timeout"变量值的意义是设定该时间段的长度，但是并非所有的平台都支持"read timeout"。struct pcap_rmtauth * auth 的作用是当用户登录到一台远程机器上时，一个指向"struct pcaprmtauth"的指针会保存必要的信息。char * errbuf 的作用是存储错误信息。

3）pcap_sendpacket()

作用：发送一个数据帧。

参数：pcap_t *p 是用来发送数据帧的网络接口；u_char *buf 是发送的数据帧的内容；int size 是发送的数据帧的大小。

返回值：–1 表示发送失败；0 表示发送成功。

4）pcap_next_ex()

作用：从某个网络接口读取或离线捕获一个数据帧。

参数：pcap_t * p 是网络接口标识；struct pcap_pkthdr ** pkt_header 是数据帧的首部信息；const u_char ** pkt_data 是数据帧的数据部分的内容。

返回值：1 表示成功读取；0 表示超过 pcap_open_live() 函数设置的时间未读取到数据帧；–1 表示发生某个错误；–2 表示离线捕获读取完毕。

4.4.2 网络通信协议并发机制实现技术

并行指两个或多个事件在同一时刻发生，并发指同一时间间隔内的两个或多个事件的发生，从时间片微观看，仍旧采用事件顺序执行。由于并发程序一般要处理列队等待、唤醒、单个事件，因此并发是一个宏观上的概念，但在微观层面，这些进程或处理都是按序列进行的，是根据序列处理完成的，只不过资源不会在某一处阻塞（因为通常是通过时间片轮转法来完成的），所以从宏观的角度来看，多个在同一时间到达的请求会在同一时间段被处理。因此，即便是同时到达的请求，也要排队等待执行，执行的先后顺序则取决于调度策略。在网络应用并发处理过程中，通过互斥量来控制多个线程对共享队列的访问，以实现多个线程的并发执行，从而提高程序运行处理效率。client 端线程并发执行示意图如图 4-10 所示。

在图 4-10 中，read_and_load 线程和 send 线程通过信号量 mutex_client_mtx 来实现对 client_queue 的互斥访问，同时使 read_and_load 线程的读取数据和 send 线程的发送数据实现并行执行。

server 端线程并发执行示意图如图 4-11 所示。receive 线程和 fetch_and_write 线程通过信号量 mutex_server_mtx 来实现对 server_queue 的互斥访问，同时使 receive 线程的接收数据和 fetch_and_write 线程从共享队列中提取的数据部分实现了操作的并行执行。receive 线程将接收的数据写入共享队列，fetch_and_write 线程将提取出的数据写入文件部分实现了操作的并行执行。

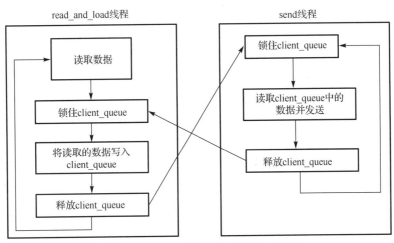

图 4-10　client 端线程并发执行示意图

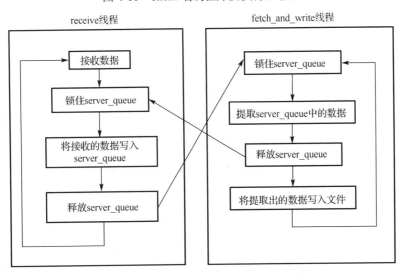

图 4-11　server 端线程并发执行示意图

4.4.3 差错控制机制实现技术

数据链路差错控制机制通过 CRC 校验码来实现。在构建数据帧时，CRC 校验码是对数据帧的首部及数据部分共同计算而获得的，所以，数据帧的构建流程如图 4-12 所示。

接收方在接收到数据帧之后，对其进行 CRC 校验，以判断在相邻节点物理链路传输过程中，数据帧是否发生比特差错，并以此达到差错控制的目的。在实现数据链路层通信协议的过程中会用到计时器。发送方在每成功发送一个数据后都会新开启一个计时器线程，计时器线程函数参数为刚发送的数据帧的序号。计时器线程在到时间后会检测对应窗口是否已被确认（接收到对方的 ACK 应答），若尚未被确认，则认为数据帧丢失，执行数据帧重发操作，而重发数据帧之前会关闭该数据帧的上一计时器，成功重发该数据帧之后会开启相应的新计时器。因此，在需要重发数据帧之前，应确保该数据帧的上一计时器线程已关闭，发送数据帧后需要开启新的计时器线程。由于线程的不可自嵌套性，因此无法在计时器线程内执行重发操作。为了解决该问题，引入 resend 线程，监控超时的计时器，以此来解决计时器的问题。计时器先通过 Sleep() 函数进行时间控制，时间控制结束后，检查该计时器对应的数据帧是否已经收到，若已经收到，则结束计时器线程；若未收到，则对重发下标 gIR.client_resend_index 的使用上锁，待 resend 线程重新发送了 gIR.client_resend_index 对应序号的数据帧后，再解锁临界资源区，并重新计时，其具体流程如图 4-13 所示。

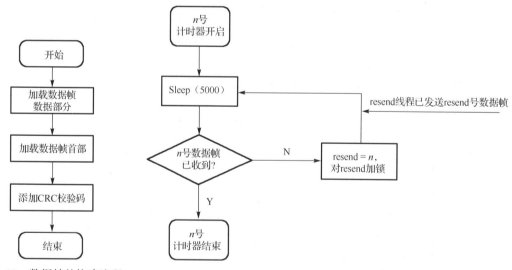

图 4-12　数据帧的构建流程　　　　　　　图 4-13　计时器线程流程

4.4.4 发送方线程与接收方线程实现技术

发送方（send）线程的主要功能是实现循环操作发送数据帧，在每次发送数据帧前，要先检查是否存在已构建的数据帧未发送，若有数据帧未发送，则立即发送，发送完成后窗口向前移动一个单位；若所有已构建的数据帧都已发送，则检查文件中是否所有的数据都已加载完成，若已加载完成，则说明数据帧已发送完成；若未加载完成，则等待 read_and_load 线程加载数据，而需要重发数据帧的工作则留给 resend 线程来执行。当需要发送应答帧时,由 send 线程负责,在达成相应条件后调用帧发送函数进行发送。send 线程工作流程如图 4-14 所示。

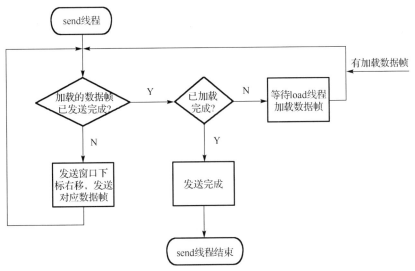

图 4-14　send 线程工作流程

数据链路层通信协议发送方源代码如下。

```
#include<stdio.h>
#include<stdlib.h>
#include<pcap.h>
#define ETHERNET_TYPE 0x0080                  //上层协议类型
#define MAX_PACKET_SIZE 1518                  //最大帧长
#define NUM_QUE 100                           //缓存队列长度
void P(int *s);                               //P 操作函数
void V(int *s);                               //V 操作函数
DWORD WINAPI read_from_file(LPVOID pM);       //文件读取进程
DWORD WINAPI send(LPVOID pM);                 //发送进程
void load_ethernet_header(u_int8_t *Buffer);  //数据帧首部构建进程
//填充数据帧,结果返回整个数据帧(包括目的 MAC 地址、源 MAC 地址、协议类型、数据字段)
int load_ethernet_data(u_int8_t *Buffer, FILE *fp);
                                              //目的 MAC 地址
u_int8_t DestinationMac[6] = { 0x20, 0x1A, 0x06, 0x21, 0xCE, 0x53 };
                                              //源 MAC 地址
u_int8_t SourceMac[6] = { 0x20, 0x1A, 0x06, 0x21, 0xCE, 0x53 };
int Mutex = 1;                                //对缓存队列访问的互斥量
int Full = 0;                                 //缓存队列溢出信号
int Empty = NUM_QUE;                          //缓存队列为空的信息,初始值设置为 100
int ReadEndFlag = 0;                          //文件读取结束标志,为 1 时标识文件读取完成
int ReadIndex = 0;                            //读进程目前在缓存队列中的读取位置(序号)
int SendIndex = 0;                            //发送进程目前在缓存队列中的发送位置(序号)
u_int8_t Pool[NUM_QUE][MAX_PACKET_SIZE] = { NULL };
                                              //缓存队列,初始值设置为 100
```

```c
struct ethernet_header
{
    u_int8_t Preamble[8];           //7字节前导码和1字节起始帧定界符
    u_int8_t DestinationMac[6];     //目的 MAC 地址
    u_int8_t SourceMac[6];          //源 MAC 地址
    u_int16_t EthernetType;         //协议类型
};

int main()
{
    /*构建线程 fread()*/
    CreateThread(NULL, 0, read_from_file, NULL, 0, NULL);
    /*构建线程 send()*/
    CreateThread(NULL, 0, send, NULL, 0, NULL);
    while (1);
}

void P(int *s)
{
    while ((*s) <= 0);
        (*s)--;
}

void V(int *s)
{
    (*s)++;
}

void load_ethernet_header(u_int8_t *Buffer)
{
    struct ethernet_header *Header = (struct ethernet_header*)Buffer;

    /*目的 MAC 地址*/
    Header->DestinationMac[0] = DestinationMac[0];
    Header->DestinationMac[1] = DestinationMac[1];
    Header->DestinationMac[2] = DestinationMac[2];
    Header->DestinationMac[3] = DestinationMac[3];
    Header->DestinationMac[4] = DestinationMac[4];
    Header->DestinationMac[5] = DestinationMac[5];
    /*源 MAC 地址*/
    Header->SourceMac[0] = SourceMac[0];
    Header->SourceMac[1] = SourceMac[1];
    Header->SourceMac[2] = SourceMac[2];
    Header->SourceMac[3] = SourceMac[3];
    Header->SourceMac[4] = SourceMac[4];
```

```c
    Header->SourceMac[5] = SourceMac[5];
    /*协议类型*/
    Header->EthernetType = ETHERNET_TYPE;
}

int load_ethernet_data(u_int8_t *Buffer, FILE *fp)
{
    int SizeOfData = 0;
    //每次最多读取1500字节作为数据帧的数据部分
    SizeOfData = fread(Buffer, 1, 1500, fp);
    //若数据帧的数据部分小于46字节,则填充若干0达到46字节
    if (SizeOfData < 46&&SizeOfData>0){
        *(Buffer + SizeOfData) = 0;
        SizeOfData++;
    }
    else if (SizeOfData == 0){
        return (sizeof(ethernet_header));
    }
    return (SizeOfData + sizeof(ethernet_header));
}

DWORD WINAPI    read_from_file(LPVOID pM)
{
    /*打开文件*/
    FILE *fp = fopen("data.txt", "r");
    if (fp == NULL){
        printf("\nthe file is opened error\n");
        system("pause");
        exit(1);
    }
    int HeaderSize = sizeof(ethernet_header);
    int flag = 0;            //设置文件读取是否完成标志
    while (ReadIndex<100){
        //取消注释,打印互斥操作过程,便于理解互斥操作
        printf("\n 前: Empty:%d  Mutex:%d\n", Empty, Mutex);
            P(&Empty);
        //取消注释,打印互斥操作过程,便于理解互斥操作
        printf("\nEmpty 后: Empty:%d  Mutex:%d\n", Empty, Mutex);
            P(&Mutex);
        //取消注释,打印互斥操作过程,便于理解互斥操作
        printf("\nMutex 后ó: 阢Empty:%d  Mutex:%d\n", Empty, Mutex);
        flag = load_ethernet_data(Pool[ReadIndex] + sizeof(ethernet_header), fp);
            if (flag == HeaderSize){
                V(&Mutex);
                V(&Full);
                break;
            }
        load_ethernet_header(Pool[ReadIndex]);
```

```c
            printf("\n生成第%d个帧n", ReadIndex);
            ReadIndex++;
            V(&Mutex);
            V(&Full);
        }
    /*文件读取完毕*/
    ReadEndFlag = 1;
}
DWORD WINAPI send(LPVOID pM)
{
    pcap_if_t *alldevs;
    pcap_if_t *d;
    pcap_t *adhandle;
    int i = 0;
    char ErrBuf[PCAP_ERRBUF_SIZE];

    /*获取本地网络接口设备列表*/
    if (pcap_findalldevs_ex(PCAP_SRC_IF_STRING, NULL, &alldevs, ErrBuf) == -1){
        printf("\nError in pcap_findalldevs_ex : %s\n", ErrBuf);
        system("pause");
        exit(1);
    }

    /*打印本地网络接口设备列表*/
    for (d = alldevs; d != NULL; d = d->next){
        printf("%d.%s", ++i, d->name);
        if (d->description)
            printf("%s\n", d->description);
        else
            printf("\nNo descriptio available.\n");
    }
    if (i == 0){
        printf("\nNo interfaces found!Make sure WinPcap is installed.\n");
        pcap_freealldevs(alldevs);
        system("pause");
        exit(1);
    }
    /*选取本次通信使用的网络接口号*/
    int INum;
    while (1){
        printf("Enter the interface nuber (1-%d):", i);
        scanf("%d", &INum);
        if (INum<1 || INum>i){
            printf("\nInterface number out of range.\n");
        }
        else break;
    }
    for (d = alldevs, i = 0; i < INum - 1; d = d->next, i++);
```

```c
    /*打开或开启已选择的网络接口*/
    if ((adhandle = pcap_open_live(
        d->name,
        65536,
        1,
        1000,
        ErrBuf
        )) == NULL){
        printf("\nUnable to open the adapter. %s is not supported by Winpcap\n",
            d->name);
        //pcap_freealldevs(alldevs);
        system("pause");
        exit(1);
    }

    /*发送数据帧*/
    while (SendIndex<=ReadIndex||ReadEndFlag==0){
        //printf("\n前: Full:%d  Mutex:%d\n", Full, Mutex);
        P(&Full);
        //printf("\nFull后: Full:%d  Mutex:%d\n", Full, Mutex);
        P(&Mutex);
        //printf("\nMutex后: Full:%d  Mutex:%d\n", Full, Mutex);
        if (pcap_sendpacket(adhandle, (const u_char*)Pool[SendIndex], 1514) != 0)
        {
            printf("\n%d.Error sending the packet:%s\n", SendIndex, pcap_geterr
                (adhandle));
        }
        else printf("Packet%d has been sent\n ", SendIndex);
        SendIndex++;
        V(&Mutex);
        V(&Empty);
    }

    /*释放网络接口设备列表*/
    pcap_freealldevs(alldevs);
    printf("\n数据帧已经发送完成\n");
    system("pause");
    exit(1);
}
```

接收方（receive）线程是数据链路层通信协议实现中较为复杂的线程。为使 client 端和 server 端集成实现全双工通信，receive 线程被设计成既具有 client 端的接收功能，又具有 server 端的接收功能。receive 线程工作流程如图 4-15 所示。

receive 线程在收到一个数据帧后，首先对数据帧进行检测，若检测错误则丢弃，若检测正确则根据接收的数据帧的长度判断其是属于 server 端发送的应答帧还是 client 端发送的数据帧。当接收的数据帧的长度为 23 字节时，为应答帧，此时 receive 线程调用其中的 client 模块，然后提取帧序号后，再检测是否达到快速重传条件，若达到快速重传条件，则重发相应帧（采用全双工通信模式），之后移动 client_queue 共享队列的发送窗口左指针。在此过程中，依据情况关闭相应计时器。当接收的数据帧的长度大于 23 字节时，说明该帧是 client 端发送的数据帧，此时 receive 线程调用其中的 server 模块提取帧序号。receive 线程先判断提取的帧序号是否在接收窗口序号范围内，若是在接收窗口序号范围内，则可以提取数据帧并存入，若不在接收窗口序号范围内，则舍弃该帧。往 server_queue 中缓存完数据帧后，检测是否有连续的帧可交付给上层协议，同时移动接收窗口左指针，直到不能交付为止，之后向发送方发送 ACK 应答。

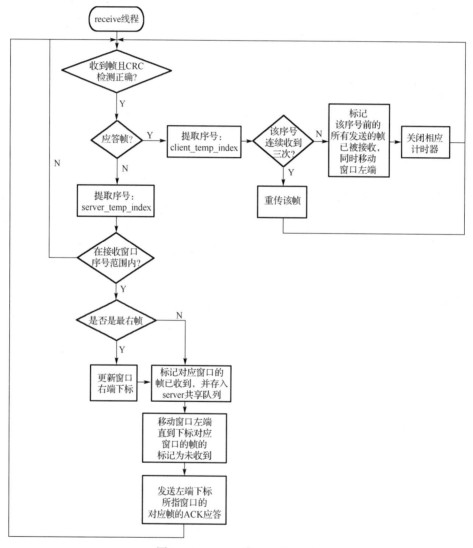

图 4-15　receive 线程工作流程

数据链路层通信协议接收方实现源代码如下。

```
#define _CRT_SECURE_NO_WARNINGS
#include <stdio.h>
#include <stdlib.h>
```

```c
#include <pcap.h>

/*宏定义部分*/
#define MAX_PACKET_SIZE 1500        //定义数据帧的数据部分的最大长度
#define NUM_QUE 100                 //定义队列长度
/*函数部分*/
void P(int *s);                     //P操作函数
void V(int *s);                     //V操作函数
//packet handler 函数说明
void packet_handler(u_char *param, const struct pcap_pkthdr *header, const
                u_char *pkt_data);
/*进程部分*/
DWORD WINAPI receive(LPVOID pM);           //接收进程,类似生产者进程
DWORD WINAPI write_to_file(LPVOID pM);     //向上层交付进程,类似消费者进程

u_int8_t AcceptDestinationMac[2][6] = { { 0xFF, 0xFF, 0xFF, 0xFF, 0xFF,
            0xFF }, { 0x20, 0x1A, 0x06, 0x21, 0xCE, 0x53 } };
u_int8_t AcceptSourceMac[2][6] = { { 0xFF, 0xFF, 0xFF, 0xFF, 0xFF, 0xFF },
            { 0x20, 0x1A, 0x06, 0x21, 0xCE, 0x53 } };

int Mutex = 1;                             //队列访问互斥量
int Full = 0;                              //队列满信号量
int Empty = NUM_QUE;                       //队列空信号量,初始值设置为100
int ReceiveIndex = 0;                      //接收进程的占用缓存区序号
int WriteIndex = 0;                        //交付进程的占用缓存区序号
int ReceiveEndFlag = 0;
u_int8_t Pool[NUM_QUE][MAX_PACKET_SIZE] = { NULL };     //队列初始化

/*以太网数据帧首部数据结构*/
struct ethernet_header
{
    u_int8_t ether_dhost[6];               //目的MAC地址
    u_int8_t ether_shost[6];               //源MAC地址
    u_int16_t ether_type;                  //协议类型
};

int main()
{   //创建receive线程
    CreateThread(NULL, 0, receive, NULL, 0, NULL);

    CreateThread(NULL, 0, write_to_file, NULL, 0, NULL);
    while (1);
}

/*P操作函数*/
void P(int *s)
{
```

```c
        while ((*s) <= 0);
        (*s)--;
}

/*V 操作函数*/
void V(int *s)
{
    (*s)++;
}

/*线程 receive()*/
DWORD WINAPI receive(LPVOID pM)
{
    pcap_if_t *alldevs;
    pcap_if_t *d;
    int inum;
    int i = 0;
    pcap_t *adhandle;
    char errbuf[PCAP_ERRBUF_SIZE];

    /*获取本地网络接口设备列表*/
    if (pcap_findalldevs_ex(PCAP_SRC_IF_STRING, NULL, &alldevs, errbuf) == -1)
    {
        fprintf(stderr, "Error in pcap_findalldevs: %s\n", errbuf);
        exit(1);
    }

    /*打印本地网络接口设备列表*/
    for (d = alldevs; d; d = d->next){
        printf("%d. %s", ++i, d->name);
        if (d->description)
            printf(" (%s)\n", d->description);
        else
            printf(" (No description available)\n");
    }
    if (i == 0){
        printf("\nNo interfaces found! Make sure WinPcap is installed.\n");
        return -1;
    }
    printf("Enter the interface number (1-%d):", i);
    scanf("%d", &inum);
    if (inum < 1 || inum > i)
    {
        printf("\nInterface number out of range.\n");
        pcap_freealldevs(alldevs);        //释放网络接口设备列表
        return -1;
```

```c
    }
    /*选择本次通信的网络接口号*/
    for (d = alldevs, i = 0; i< inum - 1; d = d->next, i++);

    /*打开或启动选择的网络接口*/
    if ((adhandle = pcap_open(d->name,       //设置网络接口初始化参数
        65536,                                //确保捕获到完整数据帧
        PCAP_OPENFLAG_PROMISCUOUS,            //将网络接口设置为混杂模式
        1000,                                 //设置读取超时时间
        NULL,                                 //远程主机验证
        errbuf                                //存储错误信息
        )) == NULL){
        fprintf(stderr, "\nUnable to open the adapter. %s is not supported"
                " by WinPcap\n", d->name);
        /*释放本地网络接口设备列表*/
        pcap_freealldevs(alldevs);
        return -1;
    }
    printf("\nlistening on %s...\n", d->description);
    /*释放本地网络接口设备列表*/
    pcap_freealldevs(alldevs);
    /*开始捕获数据帧*/
    pcap_loop(adhandle, 0, packet_handler, NULL);
    ReceiveEndFlag = 1;
}

/*每次捕获数据帧时,libpcap 都会自动调用回调函数 packet_handler()*/
void packet_handler(u_char *param, const struct pcap_pkthdr *header, const
                    u_char *pkt_data)
{
    /*将时间戳转发为可识别格式*/
    struct tm *ltime;
    char timestr[16];
    time_t local_tv_sec;

    local_tv_sec = header->ts.tv_sec;
    ltime = localtime(&local_tv_sec);
    strftime(timestr, sizeof timestr, "%H:%M:%S", ltime);

    printf("%s,%.6d len:%d\n", timestr, header->ts.tv_usec, header->len);


    u_short ethernet_type;
    struct ethernet_header *ethernet_protocol;
    u_char *mac_string;
```

```
static int packet_number = 1;
ethernet_protocol = (struct ethernet_header*)pkt_data;
int len = header->len;
int i, j;

/*判断接收到的数据帧的目的 MAC 地址是否匹配广播地址*/
int flag = 2;
for (i = 0; i < 2; i++){
    flag = 2;
    for (j = 0; j < 6; j++){
        if (ethernet_protocol->ether_dhost[j] == AcceptDestinationMac[i][j])
            continue;
        else{
            flag = i;
            break;
        }
    if (flag != 2)continue;
    else
        break;
}
if (flag != 2){
    return;
}
if (i == 0){
    printf("It's broadcasted.\n");
    return;
}

/*判断源 MAC 地址是否在可接收范围内*/
for (i = 0; i < 2; i++){
    flag = 1;
    for (j = 0; j < 6; j++){
        if (ethernet_protocol->ether_shost[j] == AcceptSourceMac[i][j])
            continue;
        else{
            flag = 0;
            break;
        }
    }
    if (flag)
        break;
}
if (flag == 0)
    return;

/*输出捕获数据帧的各字段信息*/
```

```c
        printf("\n\n----------------------------\n");
        printf("capture %d packet\n", packet_number);
        printf("capture time: %ld\n", header->ts.tv_sec);
        printf("packet length: %d\n", header->len);
        printf("-----Ethernet protocol-------\n");

        //打印数据帧中协议类型字段
        ethernet_type = ethernet_protocol->ether_type;
        printf("Ethernet type: %04x\n", ethernet_type);
        switch (ethernet_type){
            case 0x0800:printf("Upper layer protocol: IPv4\n"); break;
            case 0x0806:printf("Upper layer protocol: ARP\n"); break;
            case 0x8035:printf("Upper layer protocol: RARP\n"); break;
            case 0x814c:printf("Upper layer protocol: SNMP\n"); break;
            case 0x8137:printf("Upper layer protocol: IPX\n"); break;
            case 0x86dd:printf("Upper layer protocol: IPv6\n"); break;
            case 0x880b:printf("Upper layer protocol: PPP\n"); break;
            default:
            break;
        }

        //打印数据帧中目的 MAC 地址和源 MAC 地址
        mac_string = ethernet_protocol->ether_shost;
        printf("MAC source address: %02x:%02x:%02x:%02x:%02x:%02x\n",
            *mac_string,
            *(mac_string + 1),
            *(mac_string + 2),
            *(mac_string + 3),
            *(mac_string + 4),
            *(mac_string + 5));
        mac_string = ethernet_protocol->ether_dhost;
        printf("MAC destination address: %02x:%02x:%02x:%02x:%02x:%02x\n",
            *mac_string,
            *(mac_string + 1),
            *(mac_string + 2),
            *(mac_string + 3),
            *(mac_string + 4),
            *(mac_string + 5));

        //打印数据帧中数据部分的内容
        for (u_int8_t *p = (u_int8_t*)(pkt_data + sizeof(struct ethernet_header));
                        p != (u_int8_t*)(pkt_data + header->len); p++)
        {
```

```c
        printf("%c", *p);
    }

    /*通过互斥控制将数据帧写入接收缓存队列*/
    int k = 0;
    //取消注释，打印互斥操作过程，便于理解互斥操作
    printf("\n前°：阻 Empty:%d  Mutex:%d\n", Empty, Mutex);
    P(&Empty);
    //取消注释，打印互斥操作过程，便于理解互斥操作
    printf("\nEmpty后ó：阻 Empty:%d  Mutex:%d\n", Empty, Mutex);
    P(&Mutex);
    //取消注释，打印互斥操作过程，便于理解互斥操作
    printf("\nMutex后ó：阻 Empty:%d  Mutex:%d\n", Empty, Mutex);
    k = 0;
    for (u_int8_t *p = (u_int8_t*)(pkt_data + sizeof(struct ethernet_header));
              p != (u_int8_t*)(pkt_data + header->len); p++)
        Pool[ReceiveIndex][k++] = *p;
    //输出从接收缓存队列中读取的数据帧的数据部分
    printf("\nPool[ReceiveIndex]:%s\n", Pool[ReceiveIndex]);
    ReceiveIndex++;
    V(&Mutex);
    V(&Full);
    packet_number++;
}


//线程fwrite()
DWORD WINAPI write_to_file(LPVOID pM)
{
    /*打开文件*/
    FILE *fp = fopen("data-get.txt", "w");
    if (fp == NULL){
        printf("\nThe file is opened error\n");
        system("pause");
        exit(1);
    }
    /*通过互斥操作将数据写入文件*/
    int i = 0;
    while (WriteIndex<=ReceiveIndex||ReceiveEndFlag==0){
        //取消注释，打印互斥操作过程，以便于理解互斥操作
        printf("\n前: Full:%d  Mutex:%d\n", Full, Mutex);
        P(&Full);
        //取消注释，打印互斥操作过程，以便于理解互斥操作
```

```
        printf("\nFull 后: Full:%d  Mutex:%d\n", Full, Mutex);
        P(&Mutex);
        //取消注释,打印互斥操作过程,以便于理解互斥操作
        printf("\nMutex 后: Full:%d  Mutex:%d\n", Full, Mutex);
        //从接收缓存队列中取出数据,并写入到 data-get.txt 文件中
        fwrite(&Pool[WriteIndex], 1, sizeof(Pool[WriteIndex]), fp);
            WriteIndex++;
            V(&Mutex);
            V(&Empty);
        }
        /*数据写入文件完成*/
        fclose(fp);
        printf("\nEnd\n");
        system("pause");
        exit(1);
    }
```

4.4.5 停止-等待协议实现技术

停止-等待协议的发送窗口和接收窗口的大小均为 1。在图 4-16 中,client 端为发送方,server 端为接收方。发送方每发送一帧,发送进程便暂停发送以等待接收方的帧到达确认,接收方收到正确的帧后会发出应答帧以确认。发送方在收到接收方的应答帧之后,才会继续发送下一帧。同时,因为接收方需要知道接收到的帧是新发的还是重新发的,所以发送方发送的每一帧都需要有序号。

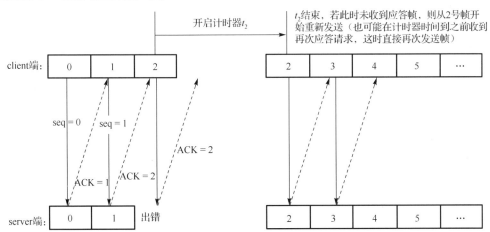

图 4-16　停止-等待协议原理示意图

在正常情况下,带差错控制的停止-等待协议,发送方发送数据帧,接收方接收数据帧并返回应答帧,发送方在收到应答帧后再继续发送下一帧。发送方在发送一帧后,停下动作,等待接收方返回的应答帧,若超过计时器时间还未收到应答帧则重发,若收到应答帧则发送下一帧。停止-等待协议的原理简单,实现也简单,send 线程和 receive 线程基本是相互触发构造的。停止-等待协议工作流程如图 4-17 所示。

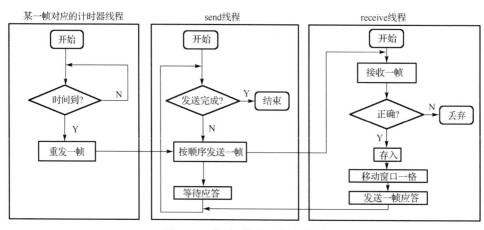

图 4-17 停止-等待工作流程图

停止-等待协议发送方实现源代码如下。

```c
/*****************************/
/*File:global.h*/
/*功能: 定义全部变量*/
/*****************************/
#ifndef _global_H_
#define _global_H_
#define _CRT_SECURE_NO_WARNINGS
#include<iostream>
#include<thread>
#include<mutex>
#include<condition_variable>
//#include<windows.h>
#include<stdlib.h>
#include<pcap.h>
#include<string.h>
using namespace std;
const int kSizeOfQueue = 100 ;
const int kSizeOfFrame = 1514;
const int kNumOfReadAndLoad = 1;            //read_and_load 线程的个数
const int kNumOfAcceptAckEthernet = 1;      //accept_ack_ethernet 线程的个数
const int kNumOfTimeClock = 1;
//帧首部的长度,因为用sizeof会被对齐影响,所以直接在此定义
const int kSizeOfHeader = 19;
const int kSizeOfCrc = 4;                   //CRC 校验码的长度
static int index_seq = 0;
static int num_of_time_clock = 0;

typedef struct ItemRepository{
    u_int8_t item_queue[kSizeOfQueue][kSizeOfFrame];
    int load_index;                         //当前待加载的帧在 queue 中的对应下标
    int send_index;                         //当前待发送的帧在 queue 中的对应下标
```

```cpp
    mutex mtx;                              //对队列访问的互斥量
    condition_variable cv_not_full;         //缓存队列未溢出的信号
    condition_variable cv_not_empty;        //缓存队列不为空的信号
}IR;

enum SeqOrAck{
seq=1,ack
};
void InitItemRepository(ItemRepository &ir);
void GenerateCrc32Table();
//kind =1:seq, kind=2:ack, 可直接输入枚举变量
int TurnToInt(u_int8_t *buffer, int kind);
#endif

/******************************/
/*File:main.cpp*/
/*功能: 停止-等待协议发送方执行入口*/
/******************************/
#include"global.h"
#include"fuc_send_ethernet.h"
#include"thread_accept_and_send.h"
#include"thread_read_and_load.h"
#include"fuc_load_frame.h"
#include"thread_time_clock.h"
int flag_read_end = 0;
int flag_load_end = 0;
int flag_send_end = 0;
IR gIR;
u_int32_t crc32_table[256];
extern thread time_clock;
int main()
{
    InitItemRepository(gIR);
    GenerateCrc32Table();

    thread read_and_load[kNumOfReadAndLoad];
    for (int i = 0; i < kNumOfReadAndLoad; ++i){
        read_and_load[i] = thread(ReadAndLoad);
    }

    thread accept_ack_ethernet[kNumOfAcceptAckEthernet];
    for (int i = 0; i < kNumOfAcceptAckEthernet; ++i){
        accept_ack_ethernet[i] = thread(AcceptAckEthernet);
    }

    for (int i = 0; i < kNumOfAcceptAckEthernet; ++i){
```

```cpp
            if (accept_ack_ethernet[i].joinable())
                accept_ack_ethernet[i].join();
        }

        for (int i = 0; i < kNumOfReadAndLoad; ++i){
            if (read_and_load[i].joinable())
                read_and_load[i].join();
        }

        for (int i = 0; i < kNumOfTimeClock; ++i){
            if (time_clock.joinable())
                time_clock.join();
        }

        system("pause");
        return 0;
}

void InitItemRepository(ItemRepository &ir){
    memset(ir.item_queue, 0, sizeof(ir.item_queue));
    ir.send_index = 0;
    ir.load_index = 0;
}
void GenerateCrc32Table(){
    int i, j;
    u_int32_t crc;
    for (i = 0; i < 256; i++)
    {
        crc = i;
        for (j = 0; j < 8; j++)
        {
            if (crc & 1)
                crc = (crc >> 1) ^ 0xEDB88320;
            else
                crc >>= 1;
        }
        crc32_table[i] = crc;
    }
}
int TurnToInt(u_int8_t *buffer, int kind)
{//kind =1:seq, kind=2:ack
    if (1 == kind){
        return
            *buffer +
            (*(buffer+1)) * 256 +
            (*(buffer+2)) * 256 * 256 +
```

```cpp
            (*(buffer+3)) * 256 * 256 * 256;
    }else
        if (2 == kind){
            return
                *buffer +
                (*(buffer + 1)) * 256 +
                (*(buffer + 2)) * 256 * 256 +
                (*(buffer + 3)) * 256 * 256 * 256;
        }
        else{
            cout << "Error:kind is error.\n" << endl;
        }
        return -1;
}


/**********************************/
/*File: fuc_load_frame.h*/
/*功能: 封装数据帧首部头文件*/
/**********************************/
#ifndef _fuc_load_frame_H_
#define _fuc_load_frame_H_
//Realtek PCIe GBE Family Controller
const u_int8_t gAcceptedSourceMac[2][6] = { { 0xFF, 0xFF, 0xFF, 0xFF, 0xFF,
       0xFF }, { 0x21,0x1A,0x06,0x21,0xCE,0x53 } };
const u_int8_t gClientMac[2][6] = { { 0xFF, 0xFF, 0xFF, 0xFF, 0xFF, 0xFF },
       { 0x20, 0x1A, 0x06, 0x21, 0xCE, 0x53 } };
/*采用小端存储, 所以要将0800改为0008, 当实际存储为低位-0800-高位, 协议栈才能读出是
  IPv4协议类型*/
const u_int16_t gEthernetType = 0x0008;
const u_int16_t gIdentify = 0x01ea;      //因为是小端存储, 所以捕获的数据帧中读取出来应该是ea01
const int kSizeOfMaxFrame = 1514;//14+3+1493+4
const int kSizeOfMinFrame = 64;  //14+46+4
enum DestinationMac{
    destinational_broadcast_mac = 1, destinational_client_mac
};
enum SourceMac{
    source_broadcast_mac = 1, source_server_mac
};
struct EthernetHeader
{
    u_int8_t Preamble[8];              //前导码
    u_int8_t destination_mac[6];       //目标MAC地址
    u_int8_t source_mac[6];            //源MAC地址
    u_int16_t ethernet_type;           //协议类型
    u_int16_t identify;
    u_int16_t len;           //整个数据帧的长度, 因为有填充, 所以将seq置于len之后
```

```cpp
        u_int8_t seq;              //用来区别其他以太网数据帧的特殊标识字段
};

/*
功能: 封装以太网数据帧的数据字段
返回值: 数据帧首部+数据部分字节大小，异常情况返回-1
void LoadFrame(u_int8_t *buffer, FILE *fp);
/*
功能: 封装以太网数据帧的首部
返回值: 执行成功返回1
*/
void LoadEthernetHeader(u_int8_t *buffer, u_int16_t len);
void LoadEthernetCrc(u_int8_t *buffer, int len);
void GenerateCrc32Table();
#endif


/*******************************/
/*File: fuc_load_frame.cpp*/
/*功能: 封装数据帧首部实现文件*/
/*******************************/
#include"global.h"
#include"fuc_load_frame.h"
extern int flag_read_end;
extern int flag_load_end;
extern u_int32_t crc32_table[256];
u_int8_t seq = 0x00;
void LoadFrame(u_int8_t *buffer, FILE *fp)
{
    memset(buffer, 0, kSizeOfFrame);

    int size_of_header;
    int size_of_data;
    int size_of_crc;

    size_of_header = kSizeOfHeader;
    size_of_crc = 4;

    /*数据部分*/
    size_of_data = fread(
        buffer + size_of_header,
        1,
        kSizeOfMaxFrame-size_of_header-size_of_crc,
```

```cpp
        fp);

    if (size_of_data < 0){//先判断数据读取是否成功
        cout << "Error:Fread in LoadFrame is failed." << endl;
        return ;
    }

    if (0 == size_of_data){//再判断数据是否已经读取完毕
        //cout << "Success:Fread has fetched complately." << endl;
        flag_read_end = 1;
        //cout << "size_of_all_header:" << size_of_all_header << endl;
        return ;
    }

    while (size_of_data < kSizeOfMinFrame -size_of_header-size_of_crc ){
        //最后判断数据大小是否符合要求
        *(buffer + size_of_header + size_of_data) = 0;
        ++size_of_data;
    }

    /*数据帧首部*/
    LoadEthernetHeader(buffer,size_of_header+size_of_data+size_of_crc);
    //在相应位置填充数据帧首部,最后CRC校验码,因为CRC校验的检测范围为数据帧首部+数据
    //部分,所以需要在确定了seq之后再加上
    LoadEthernetCrc(buffer, size_of_header + size_of_data);
    return ;
}

void LoadEthernetHeader(u_int8_t *buffer,u_int16_t len){
    struct EthernetHeader *header = (struct EthernetHeader*)buffer;

    /*目的MAC地址*/
    header->destination_mac[0] = gAcceptedSourceMac[1][0];
    header->destination_mac[1] = gAcceptedSourceMac[1][1];
    header->destination_mac[2] = gAcceptedSourceMac[1][2];
    header->destination_mac[3] = gAcceptedSourceMac[1][3];
    header->destination_mac[4] = gAcceptedSourceMac[1][4];
    header->destination_mac[5] = gAcceptedSourceMac[1][5];
    /*源MAC地址*/
    header->source_mac[0] = gClientMac[1][0];
    header->source_mac[1] = gClientMac[1][1];
    header->source_mac[2] = gClientMac[1][2];
    header->source_mac[3] = gClientMac[1][3];
    header->source_mac[4] = gClientMac[1][4];
    header->source_mac[5] = gClientMac[1][5];
```

```cpp
    /*协议类型*/
    header->ethernet_type = gEthernetType;
    /*标识*/
    header->identify = gIdentify;
    /*seq*/
    header->seq = seq;
    seq += 0x10;
    /*数据长度*/
    header->len = len;
    return ;
}
void LoadEthernetCrc(u_int8_t *buffer,int len){
    int i;
    u_int32_t crc;
    GenerateCrc32Table();

    crc = 0xffffffff;
    for (i = 0; i < len; i++)
    {
        crc = (crc >> 8) ^ crc32_table[(crc & 0xFF) ^ buffer[i]];
    }
    crc ^= 0xffffffff;
    *(u_int32_t*)(buffer + len) = crc;
}

/*********************************/
/*File:  fuc_open_adapter.h*/
/*功能：启动网络接口头文件*/
/*********************************/
#ifndef _fuc_open_adapter_H_
#define _fuc_open_adapter_H_
pcap_t *OpenAdapter();
#endif

/*********************************/
/*File:  fuc_open_adapter.cpp*/
/*功能：启动网络接口实现文件*/
/*********************************/
#include"global.h"
#include"fuc_open_adapter.h"
pcap_t *adhandle = OpenAdapter();

pcap_t *OpenAdapter(){
    pcap_if_t *alldevs;//pcap_if_t和pcap_if几乎一样
    pcap_if_t *d;
```

```
pcap_t *adhandle;
int i = 0;
char ErrBuf[PCAP_ERRBUF_SIZE];

/*获取本地网络接口设备列表*/
if (pcap_findalldevs_ex(PCAP_SRC_IF_STRING, NULL, &alldevs, ErrBuf) == -1){
    printf("\nError in pcap_findalldevs_ex : %s\n", ErrBuf);
    system("pause");
    exit(1);
}

/*打印网络接口设备列表*/
for (d = alldevs; d != NULL; d = d->next){
    printf("%d.%s", ++i, d->name);
    if (d->description)
        printf("%s\n", d->description);
    else
        printf("\nNo descriptio available.\n");
}
if (i == 0){
    printf("\nNo interfaces found!Make sure WinPcap is installed.\n");
    pcap_freealldevs(alldevs);
    system("pause");
    exit(1);
}

/*选择网络接口设备*/
int INum;
while (1){
    printf("Enter the interface nuber (1-%d):", i);
    scanf("%d", &INum);
    if (INum<1 || INum>i){
        printf("\nInterface number out of range.\n");
    }
    else break;
}
for (d = alldevs, i = 0; i < INum - 1; d = d->next, i++);

/*打开选择的网络接口设备*/
if ((adhandle = pcap_open_live(
    d->name,
    65536,
    1,
    1000,
    ErrBuf
```

```
        )) == NULL){
            printf("\nUnable to open the adapter. %s is not supported by
                Winpcap\n", d->name);
            pcap_freealldevs(alldevs);
            system("pause");
            exit(1);
        }
        else {
            pcap_freealldevs(alldevs);
            return adhandle;
        }
}
/******************************/
/*File: fuc_send_ethernet.h*/
/*功能：数据帧发送头文件*/
/******************************/
#ifndef _fuc_send_ethernet_H_
#define _fuc_send_ethernet_H_
void SendEthernet(u_int8_t seq);
#endif
/******************************/
/*File: fuc_send_ethernet.cpp*/
/*功能：数据帧发送实现文件*/
/******************************/
#include"global.h"
#include"fuc_open_adapter.h"
#include" fuc_send_ethernet.h"
#include"fuc_load_frame.h"
#include"thread_time_clock.h"
extern IR gIR;
extern u_int8_t received_ack;
extern u_int8_t sent_seq;
extern pcap_t *adhandle;
extern int flag_load_end;
extern int flag_send_end;
thread time_clock;
bool flag_over_time = false;
bool flag_arrive = false;
void SendEthernet(u_int8_t seq)
{
    unique_lock<mutex> lock(gIR.mtx);

    while (gIR.send_index == gIR.load_index){
        cout << "The shared queue is empty, SendEthernet is waiting for an
            item..."<<endl
        if (1 == flag_load_end){
```

```cpp
            cout << "Have sent finished." << endl;
            flag_send_end = 1;
            return;
        }

    (gIR.cv_not_empty).wait(lock);          //等待缓存队列不为空的信号
}

struct EthernetHeader *ethernet_header = (struct EthernetHeader*)
        (gIR.item_queue[gIR.send_index]);

    if (0 != pcap_sendpacket(adhandle, (const u_char*)gIR.item_queue[gIR.
            send_index], ethernet_header->len)){
    //数据帧最大为1514字节，不考虑CRC字段的4个字节
    cout << "Error:the errmsg of seq=" << gIR.send_index << " packet sended
            is:" << pcap_geterr(adhandle) << endl;
    flag_send_end = 1;                      //直接置1表示结束
    return;
    }
    else {
            sent_seq = ethernet_header->seq;

            if (time_clock.joinable()){
                time_clock.join();
        }
        flag_arrive = false;
        flag_over_time = false;
        time_clock = thread(TimeClock);
        cout <<"\n第"<<gIR.send_index<<"号窗口帧已经发送完毕，等待应答" <<endl;
        printf("Sent data in fuc_send_ethernet:\n%d:\n%s\n", gIR.send_
                index, gIR.item_queue[gIR.send_index]+kSizeOfHeader);
        }
        ++gIR.send_index;
        if (kSizeOfQueue == gIR.send_index)
            gIR.send_index = 0;
        (gIR.cv_not_full).notify_all();
        lock.unlock();
}

/*******************************/
/*File: thread_accept_and_send.h*/
/*功能：接收ACK应答帧并发送数据帧的头文件*/
/*******************************/
#ifndef _thread_accept_and_send_H_
#define _thread_accept_and_send_H_
void AcceptAckEthernet();
```

```cpp
//CRC 校验正确 return 1，CRC 校验错误 return -1，其他 return 0
int HeaderCheck(const u_char* pkt_data);
bool CrcCheckOut(u_int8_t *buffer, int len);

#endif

/*******************************/
/*File: thread_accept_and_send.cpp*/
/*功能：接收 ACK 应答帧并发送数据帧的实现文件*/
/*******************************/
#include"global.h"
#include"thread_accept_and_send.h"
#include"fuc_open_adapter.h"
#include"fuc_send_ethernet.h"
#include"fuc_load_frame.h"
extern IR gIR;
u_int8_t received_ack;
u_int8_t sent_seq=0x00;
extern pcap_t *adhandle;
extern int flag_send_end;
struct pcap_pkthdr *header;
const u_char *pkt_data;
extern u_int32_t crc32_table[256];
extern bool flag_arrive;          //收到为 1
extern bool flag_over_time;       //超时为 1
int header_check;
extern thread time_clock;
void AcceptAckEthernet(){

    SendEthernet(sent_seq);
    int res=0;
    while (1){
        if ((res= pcap_next_ex(adhandle, &header, &pkt_data))>=0){
        if (0 == res){            //ahandle 的超时时间到
            continue;
        }

        if (-1 == res){
          cout << "Error in thread_accept_and_send: " << pcap_geterr(adhandle);
        }

        if (true == flag_over_time){
            flag_over_time = false;
            flag_arrive = true;   //结束所有定时器
            if (0 == gIR.send_index){
            gIR.send_index = kSizeOfQueue-1;
```

```cpp
            }
            else{
           gIR.send_index--;
            }

        cout << "\n第 " << gIR.send_index << " 号窗口帧超时,将重新发送" << endl;
        SendEthernet(sent_seq);
        continue;
        }
        header_check = HeaderCheck(pkt_data);
        if (1 == header_check){
        struct EthernetHeader* received_ethernet_header = (struct
            EthernetHeader*)pkt_data;
        received_ack = received_ethernet_header->seq;   //序号的校验
        //若序号是要求的帧序号,则从共享队列取出并发送下一帧
        if (sent_seq +0x10 == received_ack){
            flag_arrive = true;
            if (time_clock.joinable()){
                time_clock.join();
            }
            cout << "收到第 "<<gIR.send_index<<" 号窗口的应答帧,正确,发送下一
                帧" << endl;
            SendEthernet(received_ack);
        }
        else{
        cout << "收到第 " << gIR.send_index << " 号窗口的应答帧,不符合要求,
                继续等待" << endl;
            }
        }
    }

    if (1 == flag_send_end){
         cout << "发送完毕" << endl;
          return;
        }
    }
}
int HeaderCheck(const u_char* pkt_data){
    //CRC 校验正确 return 1,CRC 校验错误 return -1,其他 return 0
    struct EthernetHeader* ethernet_header = (struct EthernetHeader*)pkt_data;
    /*CRC*/
    //帧首部+数据部分校验
    if (false == CrcCheckOut((u_int8_t*)pkt_data, ethernet_header->len -
         kSizeOfCrc))           return -1;
    /*协议类型校验*/
    if (ethernet_header->ethernet_type != gEthernetType){
```

```
        return 0;
    }
    /*特殊标识校验*/
    if (ethernet_header->identify != gIdentify){
        return 0;
    }

    /*目的 MAC 地址校验*/
    int i = 0, j = 0;
    //判断是否为发送到客户端对应的目的 MAC 地址
    int destination_flag = 0;
    for (i = 0; i < 2; ++i){
        for (j = 0; j < 6; ++j){
            if (ethernet_header->destination_mac[j] != gClientMac[i][j])
                break;
        }
        if (6 == j){
            destination_flag = i + 1;//flag=1:broadcast_mac  flag=2:client_mac
            break;
        }
    }
    if (0 == destination_flag){
        return 0;
    }
    if (destinational_broadcast_mac == destination_flag){
        return 0;
    }
    //数据帧目的 MAC 地址为该接收方 MAC 地址
    if (destinational_client_mac == destination_flag){
        int source_flag = 0;  //判断是否为可接收的源 MAC 地址
        for (i = 0; i < 2; ++i){
            for (j = 0; j < 6; ++j){
                if (ethernet_header->source_mac[j] != gAcceptedSourceMac[i][j])
                    break;
            }
            if (6 == j){
                source_flag = i + 1;
                break;
            }
        }
        if (0 == source_flag){
            return 0;
        }
        if (source_broadcast_mac == source_flag){
            return 0;
        }
```

```cpp
        //判断是否为可接收的服务器端的地址
        if (source_server_mac == source_flag){
            return 1;
        }
    }
    return 0;
}
bool CrcCheckOut(u_int8_t *buffer, int len){
    int i;
    u_int32_t crc;

    crc = 0xffffffff;
    for (i = 0; i < len; i++)
    {
        crc = (crc >> 8) ^ crc32_table[(crc & 0xFF) ^ buffer[i]];
    }
    crc ^= 0xffffffff;

    if (*(u_int32_t*)(buffer + len) == crc)
        return true;

    return false;
}

/********************************/
/*File: thread_read_and_load.h*/
/*功能: 读取数据并封装成数据帧的头文件*/
/********************************/
#ifndef _thread_read_and_load_H_
#define _thread_read_and_load_H_
void ReadAndLoad();
void LoadEthernetCrc(u_int8_t *buffer, int len);
void GenerateCrc32Table();
#endif
/********************************/
/*File: thread_read_and_load.cpp*/
/*功能: 读取数据并封装成数据帧的实现文件*/
/********************************/



#include"global.h"
#include"thread_read_and_load.h"
#include"fuc_load_frame.h"
extern IR gIR;
extern int flag_read_end;
extern int flag_load_end;
```

```
void ReadAndLoad(){
    FILE *fp = fopen("data.txt", "r");
    if (NULL == fp){
        cout << "File opened error."<<endl;
        exit(1);
    }

    while (1){
        unique_lock<mutex> lock(gIR.mtx);          //加锁

        while (gIR.send_index == (gIR.load_index+1)%kSizeOfFrame){
            cout << "The shared queue is full, fuc_ReadAndLoad is waiting for
                    an empty..."<<endl;
            (gIR.cv_not_full).wait(lock);          //等待收到缓存队列未溢出的信号
        }
        //装载数据帧到该共享队列的 index 位置
        LoadFrame(gIR.item_queue[gIR.load_index], fp);
        ++(gIR.load_index);
        if (kSizeOfFrame == gIR.load_index)
            gIR.load_index = 0;

        if (1 == flag_read_end)
            flag_load_end = 1;

        (gIR.cv_not_empty).notify_all();           //发出缓存队列不为空的信号

        lock.unlock() ;                            //解锁

        if (1 == flag_load_end){
            return;
        }
    }
}
/*******************************/
/*File: thread_time_clock.h*/
/*功能: 定时器头文件*/
/*******************************/
#ifndef _thread_time_clock_H_
#define _thread_time_clock_H_
void TimeClock();
#endif


/*******************************/
/*File: hread_time_clock.cpp*/
/*功能: 定时器实现文件*/
/*******************************/
```

```
#include"global.h"
#include"thread_time_clock.h"
extern bool flag_over_time;
extern bool flag_arrive;
extern IR gIR;
void TimeClock(){
    int second = 4;                         //设定超时重传时间为10s

    while (second>0){
        Sleep(1000);
        if (true == flag_arrive)return;
        second--;
    }

    if (false == flag_arrive){
        flag_over_time = true;              //超时重传时间到则重发
        return;
    }

    return;
}
```

停止-等待协议接收方实现源代码如下。

```
/******************************/
/*File: global.h*/
/*功能: 定义全局变量、数据结构和函数*/
/******************************/
#ifndef _global_H_
#define _global_H_
#define _CRT_SECURE_NO_WARNINGS
#include<iostream>
#include<mutex>
#include<pcap.h>
#include<thread>
#include<string.h>
#include<condition_variable>
using namespace std;
const int kSizeOfQueue = 100;
const int kSizeOfFrame = 1514;
const int kNumOfExtractAndWrite = 1;
const int kNumOfReceiveEthernet = 1;
const int kPcapErrbufSize = 256;
const int kSizeOfHeader = 19;
const int kSizeOfCrc = 4;
/*
const int kNumOfSendAckEthernet = 1;
```

```cpp
 */
typedef struct ItemRepository{
    u_int8_t item_queue[kSizeOfQueue][kSizeOfFrame];
    int receive_index;
    int extract_index;

    mutex mtx;
    condition_variable cv_not_full;        //缓存队列未溢出的信号
    condition_variable cv_not_empty;       //缓存队列不为空的信号
    condition_variable cv_got_seq;         //收到帧信号
}IR;
enum SeqOrAck{
    seq = 1, ack
};
void InitItemRepository(ItemRepository &ir);
void GenerateCrc32Table();
//kind =1:seq, kind=2:ack,可直接输入枚举变量
int TurnToInt(u_int8_t *buffer, int kind);
#endif
/*******************************/
/*File: main.cpp*/
/*功能: main文件、程序执行入口*/
/*******************************/
#include"global.h"
#include"thread_receive_load_ack.h"
#include"thread_extract_and_write.h"
#include"fuc_send_ack_ethernet.h"
u_int8_t received_seq=0;
u_int8_t sent_ack=0;
IR gIR;
u_int32_t crc32_table[256];
int main(){
    InitItemRepository(gIR);
    GenerateCrc32Table();

    thread receive_ethernet[kNumOfReceiveEthernet];
    for (int i = 0; i < kNumOfReceiveEthernet; ++i){
        receive_ethernet[i] = thread(ReceiveEthernet);
    }

    thread extract_and_write[kNumOfExtractAndWrite];
    for (int i = 0; i < kNumOfExtractAndWrite; ++i){
        extract_and_write[i] = thread(ExtractAndWrite);
    }

    for (int i = 0; i < kNumOfReceiveEthernet; ++i){
```

```cpp
            receive_ethernet[i].join();
        }

        for (int i = 0; i < kNumOfExtractAndWrite; ++i){
            extract_and_write[i].join();
        }

        while (1);
        return 0;
    }

    void InitItemRepository(ItemRepository &ir){
        ir.extract_index = 0;
        ir.receive_index = 0;
    }

    void GenerateCrc32Table(){
        int i, j;

        u_int32_t crc;
        for (i = 0; i < 256; i++)
        {
            crc = i;
            for (j = 0; j < 8; j++)
            {
                if (crc & 1)
                    crc = (crc >> 1) ^ 0xEDB88320;
                else
                    crc >>= 1;
            }
            crc32_table[i] = crc;
        }
    }
    //kind =1:seq, kind=2:ack
    int TurnToInt(u_int8_t *buffer, int kind){
        if (1 == kind){
            return
                *buffer +
                (*(buffer + 1)) * 256 +
                (*(buffer + 2)) * 256 * 256 +
                (*(buffer + 3)) * 256 * 256 * 256;
        }
        else
            if (2 == kind){
                return
                    *buffer +
```

```cpp
                (*(buffer + 1)) * 256 +
                (*(buffer + 2)) * 256 * 256 +
                (*(buffer + 3)) * 256 * 256 * 256;
        }
        else{
            cout << "Error:kind is error.\n" << endl;
        }
        return -1;
}
/*************************************/
/*File:fuc_open_adapter.h*/
/*功能: 打开网络接口头文件*/
/*************************************/
#ifndef _fuc_open_adapter_H_
#define _fuc_open_adapter_H_
pcap_t *OpenAdapter();
#endif


/*************************************/
/*File:fuc_open_adapter.cpp*/
/*功能: 打开网络接口实现文件*/
/*************************************/
#include"global.h"
#include"fuc_open_adapter.h"
pcap_t *adhandle = OpenAdapter();
pcap_t *OpenAdapter(){
    pcap_if_t *alldevs;              //pcap_if_t和pcap_if 几乎一样
    pcap_if_t *d;
    pcap_t *adhandle;
    int i = 0;
    char ErrBuf[PCAP_ERRBUF_SIZE];

    /*获取本地网络接口设备列表*/
    if (pcap_findalldevs_ex(PCAP_SRC_IF_STRING, NULL, &alldevs, ErrBuf) == -1){
        printf("\nError in pcap_findalldevs_ex : %s\n", ErrBuf);
        system("pause");
        exit(1);
    }

    /*打印本地网络接口设备列表*/
    for (d = alldevs; d != NULL; d = d->next){
        printf("%d.%s", ++i, d->name);
        if (d->description)
            printf("%s\n", d->description);
        else
            printf("\nNo descriptio available.\n");
```

```c
    }
    if (i == 0){
        printf("\nNo interfaces found!Make sure WinPcap is installed.\n");
        pcap_freealldevs(alldevs);
        system("pause");
        exit(1);
    }

    /*选择本次通信需要的网络接口设备*/
    int INum;
    while (1){
        printf("Enter the interface nuber (1-%d):", i);
        scanf("%d", &INum);
        if (INum<1 || INum>i){
            printf("\nInterface number out of range.\n");
        }
        else break;
    }
    for (d = alldevs, i = 0; i < INum - 1; d = d->next, i++);

    /*打开选择的网络接口设备*/
    if ((adhandle = pcap_open_live(
        d->name,
        65536,
        1,
        1000,
        ErrBuf
        )) == NULL){
        printf("\nUnable to open the adapter. %s is not supported by Winpcap\n", d->name);
        pcap_freealldevs(alldevs);
        system("pause");
        exit(1);
    }
    else {
        pcap_freealldevs(alldevs);
        return adhandle;
    }
}

/********************************/
/*File:fuc_analyze_frame.h*/
/*功能: 数据帧分析头文件*/
/********************************/
#ifndef _fuc_analyze_frame_H_
#define _fuc_analyze_frame_H_
```

```cpp
const u_int16_t gEthernetType = 0x0008;//2字节
const u_int16_t gIdentify = 0x01ea;
//自身两个地址（一个二层广播地址，一个网络接口自身的MAC地址）
const u_int8_t gServerMac[2][6] = { { 0xFF, 0xFF, 0xFF, 0xFF, 0xFF, 0xFF },
            { 0x3C, 0x77, 0xE6, 0x25, 0x91, 0x00 } };
//可接收的网络接口的MAC地址
const u_int8_t gAcceptedSourceMac[2][6] = { { 0xFF, 0xFF, 0xFF, 0xFF, 0xFF,
            0xFF }, { 0x3C, 0x77, 0xE6, 0x25, 0x91, 0x00 } };
enum DestinationMac{
    destinational_broadcast_mac = 1,destinational_server_mac
};
enum SourceMac{
    source_broadcast_mac = 1, source_client_mac
};
//定义以太网数据帧首部数据结构
struct EthernetHeader
{
    //u_int8_t Preamble[8];            //前导码
    u_int8_t destination_mac[6];       //目标MAC地址
    u_int8_t source_mac[6];            //源MAC地址
    u_int16_t ethernet_type;           //协议类型
    u_int16_t identify;
    u_int16_t len;                     //整个数据帧的长度，帧头+数据部分+CRC校验码
    u_int8_t seq;                      //用来区别其他以太网数据帧的特殊标识字段
};
#endif


/*****************************/
/*File:fuc_analyze_frame.cpp*/
/*功能：数据帧分析实现文件*/
/*****************************/
#include"global.h"
#include"fuc_open_adapter.h"
pcap_t *adhandle = OpenAdapter();
pcap_t *OpenAdapter(){
    pcap_if_t *alldevs;//pcap_if_t 和 pcap_if 几乎一样
    pcap_if_t *d;
    pcap_t *adhandle;
    int i = 0;
    char ErrBuf[PCAP_ERRBUF_SIZE];

    /*获取本地网络接口设备列表*/
    if (pcap_findalldevs_ex(PCAP_SRC_IF_STRING, NULL, &alldevs, ErrBuf) == -1){
        printf("\nError in pcap_findalldevs_ex : %s\n", ErrBuf);
        system("pause");
```

```c
        exit(1);
    }

    /*打印本地网络接口设备列表*/
    for (d = alldevs; d != NULL; d = d->next){
        printf("%d.%s", ++i, d->name);
        if (d->description)
            printf("%s\n", d->description);
        else
            printf("\nNo descriptio available.\n");
    }
    if (i == 0){
        printf("\nNo interfaces found!Make sure WinPcap is installed.\n");
        pcap_freealldevs(alldevs);
        system("pause");
        exit(1);
    }

    /*选择网络接口设备*/
    int INum;
    while (1){
        printf("Enter the interface nuber (1-%d):", i);
        scanf("%d", &INum);
        if (INum<1 || INum>i){
            printf("\nInterface number out of range.\n");
        }
        else break;
    }
    for (d = alldevs, i = 0; i < INum - 1; d = d->next, i++);

    /*打开选择的网络接口设备*/
    if ((adhandle = pcap_open_live(
        d->name,
        65536,
        1,
        1000,
        ErrBuf
        )) == NULL){
        printf("\nUnable to open the adapter. %s is not supported by Winpcap\n", d->name);
        pcap_freealldevs(alldevs);
        system("pause");
        exit(1);
    }
    else {
        pcap_freealldevs(alldevs);
        return adhandle;
```

```
        }
}

/*********************************/
/*File:fuc_send_ack_ethernet.h*/
/*功能: 构建数据帧并发送数据帧的头文件*/
/*********************************/
#ifndef _fuc_send_ack_ethernet_H_
#define _fuc_send_ack_ethernet_H_
int SendAckEthernet(u_int8_t ack, int source_index);

/*
功能: 封装以太网数据帧的首部
返回值: 执行成功返回1
*/
void LoadEthernetHeader(u_int8_t *buffer ,u_int8_t ack,int source_index);
void LoadEthernetCrc(u_int8_t *buffer, int len);


/*********************************/
/*File:fuc_send_ack_ethernet.cpp*/
/*功能: 构建数据帧并发送数据帧的实现文件*/
/*********************************/
#include"global.h"
#include"fuc_send_ack_ethernet.h"
#include"fuc_analyze_frame.h"
#include"fuc_open_adapter.h"

extern pcap_t *adhandle;
extern u_int8_t sent_ack;
extern u_int8_t sent_seq;
extern u_int32_t crc32_table[256];

int SendAckEthernet(u_int8_t ack, int source_index){
    u_int8_t ack_frame[kSizeOfHeader+kSizeOfCrc];

    /*应答帧首部*/
    LoadEthernetHeader(ack_frame,ack,source_index);

    /*应答帧CRC校验码*/
    LoadEthernetCrc(ack_frame, kSizeOfHeader);
    if (0 != pcap_sendpacket(adhandle, (const u_char*)ack_frame,
            kSizeOfHeader+kSizeOfCrc)){
        cout << "Warning:Ack sent error.\n";
        return -1;
    }
```

```cpp
    else{
        cout << "应答帧回复成功.\n";
        return 1;
    }
}

void LoadEthernetHeader(u_int8_t *buffer,u_int8_t ack, int source_index){
    struct EthernetHeader *header = (struct EthernetHeader*)buffer;

    /*目的 MAC 地址*/
    header->destination_mac[0] = gAcceptedSourceMac[source_index][0];
    header->destination_mac[1] = gAcceptedSourceMac[source_index][1];
    header->destination_mac[2] = gAcceptedSourceMac[source_index][2];
    header->destination_mac[3] = gAcceptedSourceMac[source_index][3];
    header->destination_mac[4] = gAcceptedSourceMac[source_index][4];
    header->destination_mac[5] = gAcceptedSourceMac[source_index][5];

    /*源 MAC 地址*/
    header->source_mac[0] = gServerMac[source_index][0];
    header->source_mac[1] = gServerMac[source_index][1];
    header->source_mac[2] = gServerMac[source_index][2];
    header->source_mac[3] = gServerMac[source_index][3];
    header->source_mac[4] = gServerMac[source_index][4];
    header->source_mac[5] = gServerMac[source_index][5];

    /*协议类型*/
    header->ethernet_type = gEthernetType;

    header->len = kSizeOfHeader+kSizeOfCrc;

    header->identify = gIdentify;

    header->seq = ack;
}
void LoadEthernetCrc(u_int8_t *buffer, int len){
    int i;
    GenerateCrc32Table();
    u_int32_t crc;
    crc = 0xffffffff;
    for (i = 0; i < len; i++)
    {
        crc = (crc >> 8) ^ crc32_table[(crc & 0xFF) ^ buffer[i]];
    }
    crc ^= 0xffffffff;
    *(u_int32_t*)(buffer + len) = crc;
}
```

```c
/*****************************/
/*File:thread_receive_load_ack.h*/
/*功能: 接收数据帧头文件*/
/*****************************/
#ifndef _thread_receive_load_ack_H_
#define _thread_receive_load_ack_H_
void ReceiveEthernet();
bool CrcCheckOut(u_int8_t *buffer, int len);
int HeaderCheck(const u_char* pkt_data);
#endif

/*****************************/
/*File:thread_receive_load_ack.cpp*/
/*功能: 接收数据帧实现文件*/
/*****************************/
#include"global.h"
#include"thread_receive_load_ack.h"
#include"fuc_send_ack_ethernet.h"
#include"fuc_open_adapter.h"
#include"fuc_analyze_frame.h"

extern u_int8_t received_seq;
extern u_int8_t sent_ack;
extern IR gIR;
extern pcap_t *adhandle;
extern u_int32_t crc32_table[256];
int res;
struct pcap_pkthdr *header;
const u_char *pkt_data;
char errbuf[kPcapErrbufSize];
int header_check;
int source_index = 0;

void ReceiveEthernet(){
    int res;
    while (1){
        //超时控制
        if ((res = pcap_next_ex(adhandle, &header, &pkt_data)) >= 0){
            if (0 == res)
                continue;

            header_check = HeaderCheck(pkt_data);
            if (1 == header_check){
                struct EthernetHeader* ethernet_header = (struct EthernetHeader*)pkt_data;
                received_seq = ethernet_header->seq;
```

```cpp
            if (0x00 == received_seq)
                sent_ack = received_seq;
            //若序号是要求的帧序号，则存入共享队列
            if (received_seq == sent_ack){
                cout << "\n 收到帧，符合要求，要存入队列，并返回应答帧，收到的帧长度: "
                    << ethernet_header->len << endl;
                unique_lock<mutex> lock(gIR.mtx);
                while ((gIR.receive_index + 1) % kSizeOfQueue == gIR.extract_index){
                    cout << "The shared queue is full, thread_receive_load_ack is waiting
                         for an empty...\n";
                    gIR.cv_not_full.wait(lock);
                }
                int k = 0;
                //写入共享队列，包括帧首部、数据部分和 CRC 校验码
                for (u_int8_t *p = (u_int8_t*)pkt_data; p != (u_int8_t*)(pkt_data +
                    ethernet_header->len); p++)
                    gIR.item_queue[gIR.receive_index][k++] = *p;
                printf("Loaded data in thread_receive_load_ack:\n%d:\n%s\n",
                    gIR.receive_index, gIR.item_queue[gIR.receive_index] + kSizeOfHeader);

                ++gIR.receive_index;
                if (kSizeOfQueue == gIR.receive_index)
                    gIR.receive_index = 0;

                gIR.cv_not_empty.notify_all();
                lock.unlock();

                SendAckEthernet((received_seq + 0x10), source_index);
                sent_ack = received_seq + 0x10;
            }
        }
        if (-1 == res){
            cout << "Error(in thread_accept_and_send):Reading the packets:"
                 << pcap_geterr(adhandle);
        }
    }
}
//CRC 校验正确 return 1，CRC 校验错误 return -1，其他 return 0
int HeaderCheck(const u_char* pkt_data){
    struct EthernetHeader* ethernet_header = (struct EthernetHeader*)pkt_data;
    //帧首部+数据部分校验
    if (false == CrcCheckOut((u_int8_t*)pkt_data, ethernet_header->len -
        kSizeOfCrc))        return -1;
    /*协议类型校验*/
    if (ethernet_header->ethernet_type != gEthernetType){
        return 0;
    }
```

```c
/*特殊标识校验*/
if (ethernet_header->identify != gIdentify){
    return 0;
}

/*MAC 地址校验*/
//source_index 用来指示对可接收范围内哪个客户端的 MAC 地址进行回发
int i = 0, j = 0;
//判断是否为发送到服务器端对应的 MAC 地址
int destination_flag = 0;
for (i = 0; i < 2; ++i){
    for (j = 0; j < 6; ++j){
        if (ethernet_header->destination_mac[j] != gServerMac[i][j])
            break;
    }
    if (6 == j){
        destination_flag = i + 1;//flag=1:broadcast_mac   flag=2:server_mac
        break;
    }
}
if (0 == destination_flag){
    return 0;
}
if (destinational_broadcast_mac == destination_flag){
    return 0;
}
//目的 MAC 地址为发送到服务器端地址的以太网数据帧

if (destinational_server_mac == destination_flag){
/*判断是否为可接收的地址*/
    int source_flag = 0;
    for (i = 0; i < 2; ++i){
        for (j = 0; j < 6; ++j){
        if (ethernet_header->source_mac[j] != gAcceptedSourceMac[i][j])
            break;
        }
        if (6 == j){
            source_flag = i + 1;
            source_index = i;
            break;
        }
    }
    if (0 == source_flag){
        return 0;
    }
    if (source_broadcast_mac == source_flag){
        return 0;
```

```cpp
        }
        if (source_client_mac == source_flag){
            return 1;
        }
    }
    return 0;
}

bool CrcCheckOut(u_int8_t *buffer, int len){
    int i;
    u_int32_t crc;
    crc = 0xffffffff;

    for (i = 0; i < len; i++)
    {
        crc = (crc >> 8) ^ crc32_table[(crc & 0xFF) ^ buffer[i]];
    }
    crc ^= 0xffffffff;
    if (*(u_int32_t*)(buffer + len) == crc)
        return true;
    return false;
}

/********************************/
/*File:thread_extract_and_write.h*/
/*功能：数据帧写入头文件*/
/********************************/
#ifndef _thread_extract_and_write_H_
#define _thread_extract_and_write_H_
void ExtractAndWrite();
void WriteData(u_int8_t *buffer, int len,FILE* fp);
#endif

/********************************/
/*File:thread_extract_and_write.cpp*/
/*功能：数据帧写入实现文件*/
/********************************/
#include"global.h"
#include"thread_extract_and_write.h"
#include"fuc_analyze_frame.h"
extern IR gIR;
void ExtractAndWrite(){
    FILE *fp;

    /*清空*/
    fp = fopen("data_get.txt", "w");
    if (NULL == fp){
        cout << "File opend error.\n";
```

```
        return;
    }
    fclose(fp);

    while (1){
    unique_lock<mutex> lock(gIR.mtx);
    while (gIR.extract_index  == gIR.receive_index){
    cout << "The shared queue is empty, ExtractAndWrite is waiting for an
        itme...\n";
    //等待缓存队列收到不空的信号
    (gIR.cv_not_empty).wait(lock);
    }
    struct EthernetHeader *header = (struct EthernetHeader*)(gIR.item_queue
        [gIR.extract_index]);
    cout << "in thread_extract_and_write:\nheader->len:" << header->len << endl;
    cout << "header->len-kSizeOfHeader-kSizeOfCrc:" << header->len -
        kSizeOfHeader - kSizeOfCrc << endl;
    WriteData(gIR.item_queue[gIR.extract_index]+kSizeOfHeader,header->
        len-kSizeOfHeader-kSizeOfCrc, fp);
    ++gIR.extract_index;
    if (kSizeOfQueue == gIR.extract_index)
        gIR.extract_index = 0;
    (gIR.cv_not_full).notify_all();
    lock.unlock();
    }
}

void WriteData(u_int8_t *buffer, int len,FILE* fp){
    //将追加方式写入文件
    fp = fopen("data_get.txt", "a");
    if (NULL == fp){
        cout << "File opend error.\n";
        return;
    }
    //从缓存区中读取数据并写入 data-get.txt 文件中
    fwrite(buffer, sizeof(u_int8_t), len, fp);
fclose(fp);
    return;
}
```

4.4.6 后退 N 帧协议实现技术

在后退 N 帧协议中，发送窗口大于 1，接收窗口等于 1。在发送窗口范围内，发送方可连续发送数据帧，每发送一个数据帧都要设定和启动一个计时器。如果接收到接收方的重传请求或超过计时器时间未收到相应的应答帧，则从该帧开始及其后面的帧都要重新发送。后退 N 帧协议工作原理示意图如图 4-18 所示。

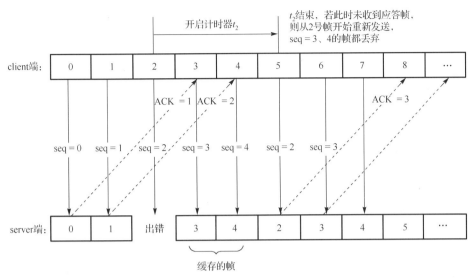

图 4-18 后退 N 帧协议工作原理示意图

在后退 N 帧协议中，发送方和接收方并非是通过相互触发达到通信同步的。发送方不停地发送数据帧，接收方每收到一个数据帧先验证数据帧的正确性。若数据帧 CRC 校验正确，则将其存储到相应的窗口对应的共享队列中，同时检查从接收窗口左端开始，一直到此时接收到的数据帧对应的窗口为止的数据帧，是否都接收到，接收方移动窗口左端到未接收到的数据帧的序号对应那个窗口位置，然后发送一个应答帧告诉接收方，到该窗口为止的所有应答帧，接收方均已正确接收，希望发送方发送该窗口对应的下一个数据帧。由于目前网络采用全双工通信模式，因此发送方一方面发送数据帧，另一方面也接收数据帧。在接收数据帧时，若连续三次收到某一窗口号的应答帧，则重发该窗口对应的数据帧，或者计时器时间到也重发该数据帧。发送方一旦开始重发某一数据帧，则该数据帧的窗口对应的后续所有数据帧都要重发。当后退 N 帧协议实现时，发送方的接收部分和接收方的接收部分均写入同一个 receive 线程中，发送方的发送部分和接收方的应答部分均写入同一个 send 线程中。后退 N 帧协议流程如图 4-19 所示。

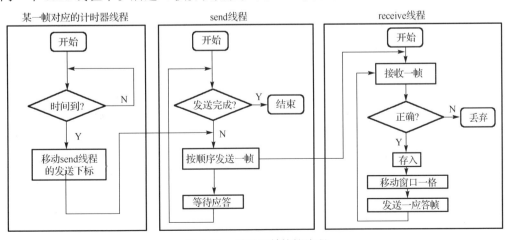

图 4-19 后退 N 帧协议流程

后退 N 帧协议发送方实现源代码如下。

```cpp
/*******************************************/
/*File:global.h*/
/*功能: 全局变量、数据结构定义头文件*/
/*******************************************/
#ifndef _global_H_
#define _global_H_
#define _CRT_SECURE_NO_WARNINGS
#include<iostream>
#include<thread>
#include<mutex>
#include<condition_variable>
//#include<windows.h>
#include<stdlib.h>
#include<pcap.h>
#include<string.h>
using namespace std;
const int kSizeOfQueue = 100 ;              //共享队列的格子大小
//数据帧的长度, 即队列的格子的大小, 不包括8字节前导码, 因为能发送的最大长度为1514 字节
const int kSizeOfFrame = 1514;
const int kNumOfReadAndLoad = 1;            //read_and_load 线程的个数
const int kNumOfSendPacket = 1;             //send_packet 线程的个数
const int kNumOfAcceptAckEthernet = 1;      //accept_ack_ethernet 线程的个数
//数据帧首部的长度, 因为用 sizeof 会被对齐影响, 所以直接在此定义
const int kSizeOfHeader = 19;
const int kSizeOfCrc = 4;                   //CRC 校验码的长度
const int kSizeOfWindows = 15;              //客户端发送窗口的大小


typedef struct ItemRepository{
    u_int8_t item_queue[kSizeOfQueue][kSizeOfFrame];
    int load_index;                         //当前待加载的帧在 queue 中的对应下标
    int send_index;                         //当前待发送的帧在 queue 中的对应下标
    int ack_index;                          //等待收到应答的帧在 queue 中的对应下标
    mutex mtx;                              //对队列访问的互斥量
    condition_variable cv_not_full;         //缓存队列未溢出的信号
    condition_variable cv_not_empty;        //缓存队列不为空的信号
    condition_variable cv_got_ack;          //确认一帧发送完成收到正确应答帧的信号
    condition_variable cv_time_used;        //时钟线程函数的控制
}IR;
enum SeqOrAck{
    seq=1,ack
};
void InitItemRepository(ItemRepository &ir);
void GenerateCrc32Table();
//kind =1:seq, kind=2:ack, 可直接输入枚举变量
```

```cpp
int TurnToInt(u_int8_t *buffer, int kind);
#endif

/***************************************/
/*File:main.cpp*/
/*功能: main 主函数执行入口程序*/
/***************************************/
#include"global.h"
#include"thread_send_ethernet.h"
#include"thread_accept_and_send.h"
#include"thread_read_and_load.h"
#include"fuc_load_frame.h"
#include"thread_time_clock.h"
int flag_read_end = 0;
int flag_load_end = 0;
int flag_send_end = 0;
IR gIR;
u_int32_t crc32_table[256];
extern thread thread_time_clock[kSizeOfQueue];
int main()
{
    InitItemRepository(gIR);
    GenerateCrc32Table();

    thread read_and_load[kNumOfReadAndLoad];
    for (int i = 0; i < kNumOfReadAndLoad; ++i){
        read_and_load[i] = thread(ReadAndLoad);
    }

    thread send_packet[kNumOfSendPacket];
    for (int i = 0; i < kNumOfSendPacket; ++i){
        send_packet[i] = thread(SendEthernet);

    }

    thread accept_ack_ethernet[kNumOfAcceptAckEthernet];
    for (int i = 0; i < kNumOfAcceptAckEthernet; ++i){
        accept_ack_ethernet[i] = thread(AcceptAckEthernet);
    }

    for (int i = 0; i < kNumOfAcceptAckEthernet; ++i){
        //当joinable()判断到当前线程未结束,可挂起当前线程(main),直到
        if (accept_ack_ethernet[i].joinable())
        //accept_ack_ethernet[i]线程运行结束后,再回来运行当前线程(main)
            accept_ack_ethernet[i].join();
```

```cpp
    }
    for (int i = 0; i < kNumOfSendPacket; ++i){
        if (send_packet[i].joinable())
            send_packet[i].join();
    }

    for (int i = 0; i < kNumOfReadAndLoad; ++i){
        //join()主要用来挂起当前函数，所以不需要考虑前面的join()会对后面的线程有影响
        if (read_and_load[i].joinable())
            //只会对当前线程（此处即main）造成阻塞
            read_and_load[i].join();
    }

    for (int i = 0; i < gIR.send_index; ++i){
        if (thread_time_clock[i].joinable())
            thread_time_clock[i].join();
    }

    while (1);
    cout << "\n发送完成" << endl;
    system("pause");
    return 0;
}
void InitItemRepository(ItemRepository &ir){
    memset(ir.item_queue, 0, sizeof(ir.item_queue));
    ir.send_index = 0;
    ir.load_index = 0;
    ir.ack_index = 0;
}

void GenerateCrc32Table(){
    int i, j;
    u_int32_t crc;
    for (i = 0; i < 256; i++)
    {
        crc = i;
        for (j = 0; j < 8; j++)
        {
            if (crc & 1)
                crc = (crc >> 1) ^ 0xEDB88320;
            else
                crc >>= 1;
        }
        crc32_table[i] = crc;
    }
```

```cpp
}
int TurnToInt(u_int8_t *buffer, int kind){//kind =1:seq, kind=2:ack
    if (1 == kind){
        return
            *buffer +
            (*(buffer+1)) * 256 +
            (*(buffer+2)) * 256 * 256 +
            (*(buffer+3)) * 256 * 256 * 256;
    }else
        if (2 == kind){
            return
                *buffer +
                (*(buffer + 1)) * 256 +
                (*(buffer + 2)) * 256 * 256 +
                (*(buffer + 3)) * 256 * 256 * 256;
        }
        else{
            cout << "Error:kind is error.\n" << endl;
        }
        return -1;
}
/***********************************/
/*File:fuc_open_adapter.h*/
/*功能: 打开网络接口头文件*/
/***********************************/
#ifndef _fuc_open_adapter_H_
#define _fuc_open_adapter_H_
pcap_t *OpenAdapter();
#endif


/***********************************/
/*File:fuc_open_adapter.cpp*/
/*功能: 打开网络接口实现文件*/
/***********************************/
#include"global.h"
#include"fuc_open_adapter.h"
pcap_t *adhandle = OpenAdapter();
pcap_t *OpenAdapter(){
    pcap_if_t *alldevs;          //pcap_if_t和pcap_if几乎一样
    pcap_if_t *d;
    pcap_t *adhandle;
    int i = 0;
    char ErrBuf[PCAP_ERRBUF_SIZE];
    /*获取本地网络接口设备列表*/
    if (pcap_findalldevs_ex(PCAP_SRC_IF_STRING, NULL, &alldevs, ErrBuf) == -1){
        printf("\nError in pcap_findalldevs_ex : %s\n", ErrBuf);
```

```c
        system("pause");
        exit(1);
    }
    /*打印网络接口设备列表*/
    for (d = alldevs; d != NULL; d = d->next){
        printf("%d.%s", ++i, d->name);
        if (d->description)
            printf("%s\n", d->description);
        else
            printf("\nNo descriptio available.\n");
    }
    if (i == 0){
        printf("\nNo interfaces found!Make sure WinPcap is installed.\n");
        pcap_freealldevs(alldevs);
        system("pause");
        exit(1);
    }

    /*选择本次通信网络接口设备*/
    int INum;
    while (1){
        printf("Enter the interface nuber (1-%d):", i);
        scanf("%d", &INum);
        if (INum<1 || INum>i){
            printf("\nInterface number out of range.\n");
        }
        else break;
    }
    for (d = alldevs, i = 0; i < INum - 1; d = d->next, i++);

    /*打开选择的网络接口设备adapter*/
    if ((adhandle = pcap_open_live(
        d->name,
        65536,
        1,
        1000,
        ErrBuf
        )) == NULL){
        printf("\nUnable to open the adapter. %s is not supported by
                Winpcap\n", d->name);
        pcap_freealldevs(alldevs);
        system("pause");
        exit(1);
    }
    else {
        pcap_freealldevs(alldevs);
```

```c
        return adhandle;
    }
}


/****************************************/
/*File:fuc_load_frame.h*/
/*功能：数据帧构造头文件*/
/****************************************/
#ifndef _fuc_load_frame_H_
#define _fuc_load_frame_H_
//可接收的源MAC地址
const u_int8_t gAcceptedSourceMac[2][6] = { { 0xFF, 0xFF, 0xFF, 0xFF, 0xFF,
        0xFF }, { 0x3C, 0x77, 0xE6, 0x25, 0x91, 0x00 } };
const u_int8_t gClientMac[2][6] = { { 0xFF, 0xFF, 0xFF, 0xFF, 0xFF, 0xFF },
        { 0x3C, 0x77, 0xE6, 0x25, 0x91, 0x00 } };
//因为计算机上是小端存储，所以要将0800改为0008，当实际存储为低位-0800-高位时，协议
//栈才能读出是IPv4协议类型
const u_int16_t gEthernetType = 0x0008;
//因为是小端存储，所以捕获的帧中读取出来应该是ea01
const u_int16_t gIdentify = 0x01ea;
const int kSizeOfMaxFrame = 1514;     //最大帧长为14+3+1493+4=1514字节
const int kSizeOfMinFrame = 64;       //最小帧长为14+46+4=64字节
enum DestinationMac{
    destinational_broadcast_mac = 1, destinational_client_mac
};
enum SourceMac{
    source_broadcast_mac = 1, source_server_mac
};

struct EthernetHeader
{
    //u_int8_t Preamble[8];          //前导码
    u_int8_t destination_mac[6];     //目标MAC地址
    u_int8_t source_mac[6];          //源MAC地址
    u_int16_t ethernet_type;         //协议类型
    u_int16_t identify;              //本程序构造的特殊帧的标识
    u_int16_t len;          //整个数据帧的长度，因为有填充，所以将seq置于len之后
    u_int8_t seq;           //用来区别其他以太网数据帧的特殊标识字段
};

/*
功能：封装以太网数据帧的数据字段
返回值：数据帧首部+数据部分字节大小，异常情况返回-1
*/
```

```cpp
void LoadFrame(u_int8_t *buffer, FILE *fp);

/*
功能: 封装以太网数据帧的首部
返回值: 执行成功返回1
*/
void LoadEthernetHeader(u_int8_t *buffer, u_int16_t len);
void LoadEthernetCrc(u_int8_t *buffer, int len);
void GenerateCrc32Table();
#endif
/******************************************/
/*File:fuc_load_frame.cpp*/
/*功能: 数据帧构造实现文件*/
/******************************************/
#include"global.h"
#include"fuc_load_frame.h"
extern int flag_read_end;
extern int flag_load_end;
extern u_int32_t crc32_table[256];
u_int8_t seq=0x00;
void LoadFrame(u_int8_t *buffer, FILE *fp)
{
    memset(buffer, 0, kSizeOfFrame);

    int size_of_header;
    int size_of_data;
    int size_of_crc;

    size_of_header = kSizeOfHeader;
    size_of_crc = 4;

    /*数据部分*/
    size_of_data = fread(
        buffer + size_of_header,
        1,
        kSizeOfMaxFrame-size_of_header-size_of_crc,
        fp);

    if (size_of_data < 0){          //先判断数据是否读取成功
        cout << "Error:Fread in LoadFrame is failed." << endl;
        return ;
    }

    if (0 == size_of_data){         //再判断数据是否已经读取完毕
        //cout << "Success:Fread has fetched complately." << endl;
```

```cpp
        flag_read_end = 1;
        //cout << "size_of_all_header:" << size_of_all_header << endl;
        return ;
    }
    //最后判断数据大小是否符合要求

    while (size_of_data < kSizeOfMinFrame -size_of_header-size_of_crc ){
        *(buffer + size_of_header + size_of_data) = 0;
        ++size_of_data;
    }

    //构造发送帧的首部
    LoadEthernetHeader(buffer,size_of_header+size_of_data+size_of_crc);
    ethernet_header->seq = seq;
    //因为CRC校验码计算包括帧首部和数据两部分，所以需要在确定了首部seq字段后，
    //才可以计算CRC校验码
    LoadEthernetCrc(buffer, size_of_header+size_of_data);
    return ;
}
void LoadEthernetHeader(u_int8_t *buffer,u_int16_t len){
    struct EthernetHeader *header = (struct EthernetHeader*)buffer;

    /*目的MAC地址*/
    header->destination_mac[0] = gAcceptedSourceMac[1][0];
    header->destination_mac[1] = gAcceptedSourceMac[1][1];
    header->destination_mac[2] = gAcceptedSourceMac[1][2];
    header->destination_mac[3] = gAcceptedSourceMac[1][3];
    header->destination_mac[4] = gAcceptedSourceMac[1][4];
    header->destination_mac[5] = gAcceptedSourceMac[1][5];

    /*源MAC地址*/
    header->source_mac[0] = gClientMac[1][0];
    header->source_mac[1] = gClientMac[1][1];
    header->source_mac[2] = gClientMac[1][2];
    header->source_mac[3] = gClientMac[1][3];
    header->source_mac[4] = gClientMac[1][4];
    header->source_mac[5] = gClientMac[1][5];
    /*协议类型*/
    header->ethernet_type = gEthernetType;
    /*标识*/
    header->identify = gIdentify;
    seq %= 0x80;
    header->seq = seq ;
    seq += 0x10;
    header->len = len;
```

```cpp
        return ;
}

void LoadEthernetCrc(u_int8_t *buffer,int len){
    int i;
    u_int32_t crc;
    GenerateCrc32Table();

    crc = 0xffffffff;
    for (i = 0; i < len; i++)
    {
        crc = (crc >> 8) ^ crc32_table[(crc & 0xFF) ^ buffer[i]];
    }
    crc ^= 0xffffffff;
    *(u_int32_t*)(buffer + len) = crc;
}


/****************************************/
/*File:thread_send_ethernet.h*/
/*功能: 发送数据帧头文件*/
/****************************************/
#ifndef _thread_send_ethernet_H_
#define _thread_send_ethernet_H_
void SendEthernet();
#endif

/****************************************/
/*File:thread_send_ethernet.cpp*/
/*功能: 发送数据帧实现文件*/
/****************************************/
#include"global.h"
#include"thread_send_ethernet.h"
#include"fuc_open_adapter.h"
#include"fuc_load_frame.h"
#include"thread_time_clock.h"

extern IR gIR;
extern u_int8_t received_ack;
extern u_int8_t sent_seq;
extern pcap_t *adhandle;
extern int flag_load_end;
extern int flag_send_end;
extern int flag_ack_end;
extern thread thread_time_clock[kSizeOfQueue];
extern bool flag_arrive[kSizeOfQueue];
```

```cpp
extern bool flag_over_time[kSizeOfQueue];

void SendEthernet()
{
    while (1){
    cout <<"\n目前窗口左端: "<< gIR.ack_index <<" 窗口右端: "<< gIR.send_index<< endl;
    unique_lock<mutex> lock(gIR.mtx);
    while (gIR.send_index == gIR.load_index){
    cout << "The shared queue is empty, SendEthernet is waiting for an item..." << endl;
        if (1 == flag_load_end){
            cout << "Have sent finished." << endl;
            flag_send_end = 1;
        }
    (gIR.cv_not_empty).wait(lock);//等待收到缓存队列不为空的信号
        if (1 == flag_ack_end){
            flag_send_end = 1;
            cout << "\nAll in thread_send_ethernet is completed.";
            return;
        }
    }
    //窗口加锁
    while (kSizeOfWindows <= (gIR.send_index - gIR.ack_index)){
        (gIR.cv_got_ack).wait(lock);
    }
    struct EthernetHeader *ethernet_header = (struct EthernetHeader*) (gIR.item_queue
            [gIR.send_index]);
    //数据帧的大小，最大为6+6+2+1500=1514字节，不考虑CRC校验码
    if (0!= pcap_sendpacket(adhandle, (const u_char*)gIR.item_queue[gIR.send_
        index], ethernet_header->len)){
        cout << "Error:the errmsg of seq=" << gIR.send_index << " packet sended is:"
            << pcap_geterr(adhandle) << endl;
        flag_send_end = 1;//直接置1表示结束
        //在这里可补充发送失败时的重发机制
    }
    else {
        printf("Sent data in thread_send_ethernet:%d:\n%s\n", gIR.send_index,
                gIR.item_queue[gIR.send_index] + kSizeOfHeader);
        /*发送成功，开启相应的计时器*/
        if (thread_time_clock[gIR.send_index].joinable()){
            thread_time_clock[gIR.send_index].join();
        }
        flag_arrive[gIR.send_index] = false;
        flag_over_time[gIR.send_index] = false;
        thread_time_clock[gIR.send_index] = thread(TimeClock,gIR.send_index);
    }
```

```cpp
        cout << "第 " << gIR.send_index << " 号窗口帧已发送成功，";
        ++gIR.send_index;
        if (kSizeOfQueue == gIR.send_index)
            gIR.send_index = 0;
        cout<<"滑动窗口右端移动至: " << gIR.send_index << " ，等待应答确认" << endl;

        (gIR.cv_not_full).notify_all();
        lock.unlock();
    }
    cout << "\nAll in thread_send_ethernet is completed.";
}

/*****************************************/
/*File:thread_read_and_load.h*/
/*功能：读取数据，并封装数据帧首部的头文件*/
/*****************************************/
#ifndef _thread_read_and_load_H_
#define _thread_read_and_load_H_
void ReadAndLoad();
void LoadEthernetCrc(u_int8_t *buffer, int len);
void GenerateCrc32Table();
#endif


/*****************************************/
/*File:thread_read_and_load.cpp*/
/*功能：读取数据，并封装数据帧首部的实现文件*/
/*****************************************/
#include"global.h"
#include"thread_read_and_load.h"
#include"fuc_load_frame.h"
extern IR gIR;
extern int flag_read_end;
extern int flag_load_end;

void ReadAndLoad(){
FILE *fp = fopen("data.txt", "r");
if (NULL == fp){
    cout << "File opened error."<<endl;
    exit(1);
}

while (flag_load_end!=1){
unique_lock<mutex> lock(gIR.mtx);    //加锁
while (gIR.send_index == (gIR.load_index+1)%kSizeOfQueue){
cout << "The shared queue is full, fuc_ReadAndLoad is waiting for an
        empty..."<<endl;
```

```
           (gIR.cv_not_full).wait(lock);           //等待收到缓存队列未溢出的信号
    }
    LoadFrame(gIR.item_queue[gIR.load_index], fp);//装载数据帧到该共享队列的index位置
    if (1 == flag_read_end){
        flag_load_end = 1;
    }
    else{
        ++(gIR.load_index);
        if (kSizeOfQueue == gIR.load_index){
        gIR.load_index = 0;
    }
    }
    (gIR.cv_not_empty).notify_all();        //发出不为空的信号
    lock.unlock();                          //解锁
    }
    cout << "\nAll in thread_read_and_load is completed.";
}


/***************************************/
/*File:thread_accept_and_send.h*/
/*功能: 帧首部检测头文件*/
/***************************************/
#ifndef _thread_accept_and_send_H_
#define _thread_accept_and_send_H_
void AcceptAckEthernet();
//CRC校验正确return 1, CRC校验错误return -1, 其他return 0
int HeaderCheck(const u_char* pkt_data);
bool CrcCheckOut(u_int8_t *buffer, int len);
#endif


/***************************************/
/*File:thread_accept_and_send.cpp*/
/*功能: 帧首部检测实现文件*/
/***************************************/
#include"global.h"
#include"thread_accept_and_send.h"
#include"fuc_open_adapter.h"
#include"thread_send_ethernet.h"
#include"fuc_load_frame.h"

extern IR gIR;
u_int8_t received_ack;
extern pcap_t *adhandle;
extern int flag_send_end;
int flag_ack_end;
```

```cpp
struct pcap_pkthdr *header;
const u_char *pkt_data;
extern u_int32_t crc32_table[256];
extern bool flag_arrive[kSizeOfQueue];
extern bool flag_over_time[kSizeOfQueue];

void AcceptAckEthernet(){
SendEthernet(sent_seq);
int res = 0;
while (1 != flag_ack_end){
unique_lock<mutex> lock(gIR.mtx);
if (true == flag_over_time[gIR.ack_index]){  //若超时时间到则重发
gIR.send_index = gIR.ack_index;
}
if ((res = pcap_next_ex(adhandle, &header, &pkt_data)) >= 0){
    if (0 == res)                              //ahandle 的超时时间到
    continue;
    int header_check;
    header_check = HeaderCheck(pkt_data);
    if (1 == header_check){
    struct EthernetHeader* ack_ethernet_header = (struct EthernetHeader*)
        pkt_data;
    received_ack = ack_ethernet_header->seq; //帧序号校验
    struct EthernetHeader *ethernet_header = (struct EthernetHeader*)
        (gIR.item_queue[gIR.ack_index]);
    //若帧序号是要求的序号,则从共享队列取出该帧并发送下一帧
    if ((ethernet_header->seq+0x10) == received_ack){
        flag_arrive[gIR.ack_index] = true;     //该帧到达,控制计时器结束
        cout << "收到的第 " << gIR.ack_index << " 号帧符合要求";
        gIR.ack_index++;
        if (kSizeOfQueue == gIR.ack_index){
            gIR.ack_index = 0;
        }
        cout<<",滑动窗口左端右移至:" << gIR.ack_index << endl;
    }
    else{
        cout <<"第 "<<gIR.ack_index<<" 号窗口的应答帧不符合要求,等待超时重传" << endl;
    }
    }
    else if (-1 == header_check){

    //CRC 校验码错误导致重传

    }
    else {
    continue;
```

```cpp
    }

    if (-1 == res){
        cout << "Error(in thread_accept_and_send):Reading the packets:"
            << pcap_geterr(adhandle);
    }
        if (gIR.ack_index == gIR.load_index){   //检测是否收到了全部应答帧
            flag_ack_end = 1;
            cout << "\nAll in thread_accept_and_send is completed.";
            (gIR.cv_not_empty).notify_all();    //发出不为空的信号
            return;
            }
        }
    }
}
//CRC校验正确return 1，CRC校验错误return -1，其他return 0
int HeaderCheck(const u_char* pkt_data){
    struct EthernetHeader* ethernet_header = (struct EthernetHeader*)pkt_data;
    /*CRC校验，帧首部+数据部分校验*/
        if (false == CrcCheckOut((u_int8_t*)pkt_data, ethernet_header->len
            - kSizeOfCrc))
            return -1;
    /*特殊标识字段校验*/
        if (ethernet_header->identify != gIdentify){
            return 0;
    }
    /*协议类型校验*/
    if (ethernet_header->ethernet_type != gEthernetType){
        return 0;
    }
    /*MAC地址校验*/
    int i = 0, j = 0;
    //判断是否为发送到本客户端对应的MAC地址
    int destination_flag = 0;
    for (i = 0; i < 2; ++i){
    for (j = 0; j < 6; ++j){
        if (ethernet_header->destination_mac[j] != gClientMac[i][j])
            break;
        }
        if (6 == j){
            destination_flag = i + 1;//flag=1（广播地址），flag=2（单播地址）
            break;
        }
    }
    if (0 == destination_flag){
```

```c
        return 0;
    }
    if (destinational_broadcast_mac == destination_flag){
        return 0;
    }
    //是目的 MAC 地址为发送到客户端地址的以太网数据帧
    if (destinational_client_mac == destination_flag){
        int source_flag = 0;          //判断是否为可接收的源 MAC 地址
        for (i = 0; i < 2; ++i){
            for (j = 0; j < 6; ++j){
                if (ethernet_header->source_mac[j] != gAcceptedSourceMac[i][j])
                    break;
            }
            if (6 == j){
                source_flag = i + 1;
                break;
            }
        }
        if (0 == source_flag){
            return 0;
        }
        if (source_broadcast_mac == source_flag){
            return 0;
        }
        //判断是否为可接收的服务器端的地址
        if (source_server_mac == source_flag){
            return 1;
        }
    }
    return 0;
}

bool CrcCheckOut(u_int8_t *buffer, int len){
    int i;
    u_int32_t crc;

    crc = 0xffffffff;
    for (i = 0; i < len; i++)
    {
        crc = (crc >> 8) ^ crc32_table[(crc & 0xFF) ^ buffer[i]];
    }
    crc ^= 0xffffffff;

    if (*(u_int32_t*)(buffer + len) == crc)
        return true;
```

```cpp
            return false;
}


/***************************************/
/*File:thread_time_clock.h*/
/*功能: 定时器头文件*/
/***************************************/
#ifndef _thread_time_clock_H_
#define _thread_time_clock_H_
void TimeClock(int index);
#endif


/***************************************/
/*File:thread_time_clock.cpp*/
/*功能: 定时器实现文件*/
/***************************************/
#include"global.h"
#include"thread_time_clock.h"

bool flag_arrive[kSizeOfQueue];
bool flag_over_time[kSizeOfQueue];
thread thread_time_clock[kSizeOfQueue];
extern IR gIR;

void TimeClock(int index){
    int seconds = 5;//超时时间设为5s
    while (seconds > 0){
        if (true == flag_arrive[index])return;
        Sleep(1000);
        seconds--;
    }
    unique_lock<mutex> lock(gIR.mtx);
    if (0 == seconds){
        flag_over_time[index] = true;

        int i;
        i = index+1;
        while (i != gIR.send_index){//关闭其他计时器
            flag_arrive[i] = true;
            i++;
            if (kSizeOfQueue == i)i = 0;
        }

        gIR.send_index = gIR.ack_index;
        cout << "\n第 " << index << " 号窗口帧超时，准备由此处开始重发" << endl;
```

```
            //cout << "Come from:" << index << endl;
            //gIR.send_index = index;

            if (gIR.send_index != gIR.load_index){

                (gIR.cv_not_empty).notify_all(); //发出缓存队列不为空的信号
            }
            return;
        }
        return;
}
```

后退 N 帧协议接收方实现源代码如下。

```
/******************************/
/*File:global.h*/
/*功能: 全局变量、数据结构定义头文件*/
/******************************/
#ifndef _global_H_
#define _global_H_
#define _CRT_SECURE_NO_WARNINGS
#include<iostream>
#include<mutex>
#include<pcap.h>
#include<thread>
#include<string.h>
#include<condition_variable>
using namespace std;

const int kSizeOfQueue = 100;
const int kSizeOfFrame = 1514;
const int kNumOfExtractAndWrite = 1;
const int kNumOfReceiveEthernet = 1;
const int kPcapErrbufSize = 256;
const int kSizeOfHeader = 19;
const int kSizeOfCrc = 4;
const int kNumOfSendAckEthernet = 1;


typedef struct ItemRepository{
    u_int8_t item_queue[kSizeOfQueue][kSizeOfFrame];
    int receive_index;
    int extract_index;
    int ack_index;

    mutex mtx;
    condition_variable cv_not_full;            //缓存队列未溢出信号
```

```cpp
    condition_variable cv_not_empty;        //缓存队列不为空的信号
    condition_variable cv_got_seq;          //收到帧信号
}IR;

enum SeqOrAck{
    seq = 1, ack
};

void InitItemRepository(ItemRepository &ir);

void GenerateCrc32Table();
//kind =1:seq，kind=2:ack，可直接输入枚举变量
int TurnToInt(u_int8_t *buffer, int kind);
#endif
/************************************/
/*File:main.cpp*/
/*功能: main 主函数执行入口实现文件*/
/************************************/
#include"global.h"
#include"thread_receive_load_ack.h"
#include"thread_extract_and_write.h"
#include"fuc_send_ack_ethernet.h"
u_int8_t received_seq=0;
u_int8_t sent_ack=0;
IR gIR;
u_int32_t crc32_table[256];
int main(){
    InitItemRepository(gIR);
    GenerateCrc32Table();

    thread receive_ethernet[kNumOfReceiveEthernet];
    for (int i = 0; i < kNumOfReceiveEthernet; ++i){
        receive_ethernet[i] = thread(ReceiveEthernet);
    }

    thread extract_and_write[kNumOfExtractAndWrite];
    for (int i = 0; i < kNumOfExtractAndWrite; ++i){
        extract_and_write[i] = thread(ExtractAndWrite);
    }

    for (int i = 0; i < kNumOfReceiveEthernet; ++i){
        receive_ethernet[i].join();
    }

    for (int i = 0; i < kNumOfExtractAndWrite; ++i){
        extract_and_write[i].join();
```

```cpp
    }
    while (1);
    return 0;
}

void InitItemRepository(ItemRepository &ir){
    ir.extract_index = 0;
    ir.receive_index = 0;
    ir.ack_index = 0;
}
void GenerateCrc32Table(){
    int i, j;

    u_int32_t crc;
    for (i = 0; i < 256; i++)
    {
        crc = i;
        for (j = 0; j < 8; j++)
        {
            if (crc & 1)
                crc = (crc >> 1) ^ 0xEDB88320;
            else
                crc >>= 1;
        }
        crc32_table[i] = crc;
    }
}
int TurnToInt(u_int8_t *buffer, int kind){//kind =1:seq   kind=2:ack
    if (1 == kind){
        return
            *buffer +
            (*(buffer + 1)) * 256 +
            (*(buffer + 2)) * 256 * 256 +
            (*(buffer + 3)) * 256 * 256 * 256;
    }
    else
        if (2 == kind){
            return
                *buffer +
                (*(buffer + 1)) * 256 +
                (*(buffer + 2)) * 256 * 256 +
                (*(buffer + 3)) * 256 * 256 * 256;
        }
        else{
            cout << "Error:kind is error.\n" << endl;
```

```
        }
        return -1;
}


/*******************************/
/*File:fuc_open_adapter.h        */
/*功能: 打开网络接口头文件*/
/*******************************/
#ifndef _fuc_open_adapter_H_
#define _fuc_open_adapter_H_
pcap_t *OpenAdapter();
#endif
/*******************************/
/*File:fuc_open_adapter.cpp*/
/*功能: 打开网络接口实现文件*/
/*******************************/
#include"global.h"
#include"fuc_open_adapter.h"
pcap_t *adhandle = OpenAdapter();
pcap_t *OpenAdapter(){
    pcap_if_t *alldevs;//pcap_if_t和pcap_if 几乎一样
    pcap_if_t *d;
    pcap_t *adhandle;
    int i = 0;
    char ErrBuf[PCAP_ERRBUF_SIZE];

    /*获取本地网络接口设备列表*/
    if (pcap_findalldevs_ex(PCAP_SRC_IF_STRING, NULL, &alldevs, ErrBuf) == -1){
        printf("\nError in pcap_findalldevs_ex : %s\n", ErrBuf);
        system("pause");
        exit(1);
    }

    /*打印网络接口设备列表*/
    for (d = alldevs; d != NULL; d = d->next){
        printf("%d.%s", ++i, d->name);
        if (d->description)
            printf("%s\n", d->description);
        else
            printf("\nNo descriptio available.\n");
    }
    if (i == 0){
        printf("\nNo interfaces found!Make sure WinPcap is installed.\n");
        pcap_freealldevs(alldevs);
        system("pause");
```

```c
        exit(1);
    }

    /*选择本次通信网络接口*/
    int INum;
    while (1){
        printf("Enter the interface nuber (1-%d):", i);
        scanf("%d", &INum);
        if (INum<1 || INum>i){
            printf("\nInterface number out of range.\n");
        }
        else break;
    }
    for (d = alldevs, i = 0; i < INum - 1; d = d->next, i++);

    /*打开选择的网络接口 adapter*/
    if ((adhandle = pcap_open_live(
        d->name,
        65536,
        1,
        1000,
        ErrBuf
        )) == NULL){
        printf("\nUnable to open the adapter. %s is not supported by Winpcap\n", d->name);
        pcap_freealldevs(alldevs);
        system("pause");
        exit(1);
    }
    else {
        pcap_freealldevs(alldevs);
        return adhandle;
    }
}


/*******************************************/
/*File:fuc_send_ack_ethernet.h*/
/*功能: 发送ACK应答帧头文件*/
/*******************************************/
#ifndef _fuc_send_ack_ethernet_H_
#define _fuc_send_ack_ethernet_H_

int SendAckEthernet(u_int8_t ack, int source_index);

/*
功能: 封装以太网数据帧的首部
```

返回值：执行成功返回 1
*/
```cpp
void LoadEthernetHeader(u_int8_t *buffer ,u_int8_t ack,int source_index);
void LoadEthernetCrc(u_int8_t *buffer, int len);
#endif
```
/***/
/*File:fuc_send_ack_ethernet.cpp*/
/*功能：发送 ACK 应答帧实现文件*/
/***/
```cpp
#include"global.h"
#include"fuc_send_ack_ethernet.h"
#include"fuc_analyze_frame.h"
#include"fuc_open_adapter.h"
extern pcap_t *adhandle;
extern u_int8_t sent_ack;
extern u_int8_t sent_seq;
extern u_int32_t crc32_table[256];
int SendAckEthernet(u_int8_t ack, int source_index){
    u_int8_t ack_frame[kSizeOfHeader+kSizeOfCrc];

    /*ACK 应答帧首部*/
    LoadEthernetHeader(ack_frame,ack,source_index);
    sent_ack = ack;

    /*ACK 应答帧 CRC 校验码*/
    LoadEthernetCrc(ack_frame, kSizeOfHeader);

    if (0 != pcap_sendpacket(adhandle, (const u_char*)ack_frame,
            kSizeOfHeader+kSizeOfCrc)){
        cout << "Warning:Ack sent error.\n";
        return -1;
    }
    else{
        cout << "Success:Ack sent successfully.\n";
        return 1;
    }
}

void LoadEthernetHeader(u_int8_t *buffer,u_int8_t ack, int source_index){
    struct EthernetHeader *header = (struct EthernetHeader*)buffer;
    /*目的 MAC 地址*/
    header->destination_mac[0] = gAcceptedSourceMac[source_index][0];
    header->destination_mac[1] = gAcceptedSourceMac[source_index][1];
    header->destination_mac[2] = gAcceptedSourceMac[source_index][2];
    header->destination_mac[3] = gAcceptedSourceMac[source_index][3];
```

```c
    header->destination_mac[4] = gAcceptedSourceMac[source_index][4];
    header->destination_mac[5] = gAcceptedSourceMac[source_index][5];

    /*源MAC地址*/
    header->source_mac[0] = gServerMac[source_index][0];
    header->source_mac[1] = gServerMac[source_index][1];
    header->source_mac[2] = gServerMac[source_index][2];
    header->source_mac[3] = gServerMac[source_index][3];
    header->source_mac[4] = gServerMac[source_index][4];
    header->source_mac[5] = gServerMac[source_index][5];

    /*协议类型*/
    header->ethernet_type = gEthernetType;

    header->len = kSizeOfHeader+kSizeOfCrc;

    header->identify = gIdentify;

    header->seq = ack;
}

void LoadEthernetCrc(u_int8_t *buffer, int len){
    int i;
    GenerateCrc32Table();
    u_int32_t crc;
    crc = 0xffffffff;
    for (i = 0; i < len; i++)
    {
        crc = (crc >> 8) ^ crc32_table[(crc & 0xFF) ^ buffer[i]];
    }
    crc ^= 0xffffffff;
    *(u_int32_t*)(buffer + len) = crc;
}


/*************************************/
/*File:thread_receive_load_ack.h*/
/*功能：接收数据帧并检测其首部头文件*/
/*************************************/
#ifndef _thread_receive_load_ack_H_
#define _thread_receive_load_ack_H_
void ReceiveEthernet();
//CRC校验正确return 1，CRC校验错误return -1,其他return 0
int HeaderCheck(const u_char* pkt_data);
bool CrcCheckOut(u_int8_t *buffer, int len);
#endif
```

```cpp
/****************************************/
/*File:thread_receive_load_ack.cpp*/
/*功能：接收数据帧并检测其首部实现文件*/
/****************************************/
#include"global.h"
#include"thread_receive_load_ack.h"
#include"fuc_send_ack_ethernet.h"
#include"fuc_open_adapter.h"
#include"fuc_analyze_frame.h"

extern u_int8_t received_seq;
extern u_int8_t sent_ack;
extern IR gIR;
extern pcap_t *adhandle;
extern u_int32_t crc32_table[256];
int res;
int source_index = 0;
struct pcap_pkthdr *header;
const u_char *pkt_data;
char errbuf[kPcapErrbufSize];

void ReceiveEthernet(){
    int res;
    while (1){
    //超时相关控制
    if ((res = pcap_next_ex(adhandle, &header, &pkt_data)) >= 0){
    if (0 == res){
       continue;
    }

    if (-1 == res){
    cout << "Error(in thread_accept_and_send):Reading the packets:"
         << pcap_geterr(adhandle);
    }
    int header_check;
    header_check = HeaderCheck(pkt_data);
    if (1 == header_check){
    struct EthernetHeader* ethernet_header = (struct EthernetHeader*)
             pkt_data;
    received_seq = ethernet_header->seq;
    //当收到的 seq 为 0x00 时，表示对方是首帧
    if (0x00 == received_seq)
    //由于非首帧的 seq 也可能是 0x00，因此这是一个隐患，只有用其他字节位表示才能解决
    sent_ack = received_seq;
    //若帧序号是要求的序号，则存入共享队列
```

```cpp
        if (sent_ack == received_seq){
        cout << "\n收到帧，符合要求，要存入队列并返回应答帧,收到的帧长度: "
            << ethernet_header->len << endl;
        unique_lock<mutex> lock(gIR.mtx);
        while ((gIR.receive_index + 1) % kSizeOfQueue == gIR.extract_index){
            cout << "The shared queue is full, thread_receive_load_ack is waiting
                for an empty...\n";
            gIR.cv_not_full.wait(lock);
        }
        int k = 0;
        //写入共享队列，包括帧首部和数据部分，以及CRC校验码
        for (u_int8_t *p = (u_int8_t*)pkt_data; p != (u_int8_t*)(pkt_data +
            ethernet_header->len); p++)
        gIR.item_queue[gIR.receive_index][k++] = *p;
        printf("Loaded data in thread_receive_load_ack:\n%d:\n%s\n",
            gIR.receive_index, gIR.item_queue[gIR.receive_index] + kSizeOfHeader);

        ++gIR.receive_index;
        if (kSizeOfQueue == gIR.receive_index){
            gIR.receive_index = 0;
        }

        ++gIR.ack_index;
        if (kSizeOfQueue == gIR.ack_index){
            gIR.ack_index = 0;
        }

        gIR.cv_not_empty.notify_all();
        lock.unlock();

        SendAckEthernet(received_seq + 0x10, source_index);
    }
        else{
        //如果不是要求的ACK应答帧，则向对方发送ACK应答
        SendAckEthernet(sent_ack, source_index);                }
        }
        else if (-1 == header_check){
            continue;
            //要求重新发送数据帧

            }


        }
    }
}
```

```c
//CRC校验正确return 1，CRC校验错误return -1，其他return 0
int HeaderCheck(const u_char* pkt_data){
    struct EthernetHeader* ethernet_header = (struct EthernetHeader*)pkt_data;
    /*CRC校验*/
    if (false == CrcCheckOut((u_int8_t*)pkt_data, ethernet_header->len -
        kSizeOfCrc))//帧首部+数据部分校验
        return -1;

    /*特殊标识字段校验*/
    if (ethernet_header->identify != gIdentify){
        return 0;
    }
    /*协议类型校验*/
    if (ethernet_header->ethernet_type != gEthernetType){
        return 0;
    }

    /*MAC地址校验*/
    int i = 0, j = 0;//source_index用来指示对可接收范围内的哪个客户端MAC地址进行回发

    int destination_flag = 0;//判断是否为发送到服务器端对应的MAC地址
    for (i = 0; i < 2; ++i){
        for (j = 0; j < 6; ++j){
            if (ethernet_header->destination_mac[j] != gServerMac[i][j])
                break;
        }
        if (6 == j){
            destination_flag = i + 1;//flag=1:broadcast_mac, flag=2:server_mac
            break;
        }
    }
    if (0 == destination_flag){
        return 0;
    }
    if (destinational_broadcast_mac == destination_flag){
        return 0;
    }
    //目的MAC地址为发送到服务器端地址的以太网数据帧
    if (destinational_server_mac == destination_flag){
        /*判断是否为可接收的地址*/
        int source_flag = 0;
        for (i = 0; i < 2; ++i){
            for (j = 0; j < 6; ++j){
                if (ethernet_header->source_mac[j] != gAcceptedSourceMac[i][j])
                    break;
```

```
            }
            if (6 == j){
                source_flag = i + 1;
                source_index = i;
                break;
            }
        }
        if (0 == source_flag){
            return 0;
        }
        if (source_broadcast_mac == source_flag){
            return 0;
        }
        if (source_client_mac == source_flag){
            return 1;
        }
    }
    return 0;
}

bool CrcCheckOut(u_int8_t *buffer, int len){
    int i;
    u_int32_t crc;
    crc = 0xffffffff;

    for (i = 0; i < len; i++)
    {
        crc = (crc >> 8) ^ crc32_table[(crc & 0xFF) ^ buffer[i]];
    }
    crc ^= 0xffffffff;
    if (*(u_int32_t*)(buffer + len) == crc)
        return true;
    return false;
}



/*******************************************/
/*File:thread_extract_and_write.h*/
/*功能: 获取数据帧数据部分并写入文件头文件*/
/*******************************************/
#ifndef _thread_extract_and_write_H_
#define _thread_extract_and_write_H_

void ExtractAndWrite();
void WriteData(u_int8_t *buffer, int len,FILE* fp);
#endif
```

```cpp
/*****************************************/
/*File:thread_extract_and_write.cpp*/
/*功能：获取数据帧数据部分并写入文件实现文件*/
/*****************************************/

#include"global.h"
#include"thread_extract_and_write.h"
#include"fuc_analyze_frame.h"

extern IR gIR;

void ExtractAndWrite(){
    FILE *fp;

    /*打开文件*/
    fp = fopen("data_get.txt", "w");
    if (NULL == fp){
        cout << "File opend error.\n";
        return;
    }
    fclose(fp);

    while (1){
        unique_lock<mutex> lock(gIR.mtx);
        while (gIR.extract_index  == gIR.receive_index){
            cout << "The shared queue is empty, ExtractAndWrite is waiting
                    for an itme...\n";
            (gIR.cv_not_empty).wait(lock);//等待收到缓存队列不为空的信号
        }

        struct EthernetHeader *header = (struct EthernetHeader*)(gIR.item_
                                    queue[gIR.extract_index]);
        //cout << "in thread_extract_and_write:\nheader->len:"
        //      << header->len << endl;
        //cout << "header->len-kSizeOfHeader-kSizeOfCrc:" << header->len -
        //      kSizeOfHeader - kSizeOfCrc << endl;
        WriteData(gIR.item_queue[gIR.extract_index]+kSizeOfHeader,
            header->len-kSizeOfHeader-kSizeOfCrc, fp);
        ++gIR.extract_index;

        if (kSizeOfQueue == gIR.extract_index)
            gIR.extract_index = 0;

        (gIR.cv_not_full).notify_all();
        lock.unlock();
    }
```

```
}

void WriteData(u_int8_t *buffer, int len,FILE* fp){

//将追加方式写入文件
fp = fopen("data_get.txt", "a");
if (NULL == fp){
    cout << "File opend error.\n";
    return;
}
//从缓存区中取出数据写入data-get.txt文件
    fwrite(buffer, sizeof(u_int8_t), len, fp);
    fclose(fp);

    return;
}
```

4.4.7 选择重传协议实现技术

在选择重传协议中，发送窗口和接收窗口均大于 1。选择重传协议与后退 N 帧协议不同的地方：当接收方发现接收的数据帧出现乱序时，只要数据帧的序号在接收窗口内，则不丢弃该数据帧，而是将该数据帧缓存下来；发送方每发送一个数据帧，便启动一个重发定时器，如果重发定时器超时且该数据帧对应的应答还没有收到，则发送方认为该数据帧出错，可能是比特差错，也有可能是传输差错，此时发送方仅发送出错的数据帧。选择重传协议工作原理示意图如图 4-20 所示。

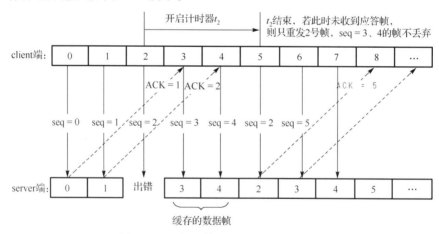

图 4-20 选择重传协议工作原理示意图

在选择重传协议实现中，当某一数据帧的重发计时器到时间时，发送方只重发出错的数据帧，而不是重发该数据帧后的所有帧。当连续收到三次同一应答帧时，采用快速重传协议重发相应数据帧。当接收方接收到正确数据帧后，只要接收窗口数据缓存连续，则一并交付相应数据帧给上层协议，并向对方发送 ACK 应答，发送方收到相应确认，则知道该数据帧及其前所有数据帧都已被正确接收。选择重传协议流程如图 4-21 所示。

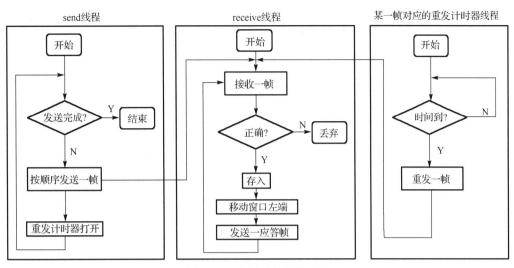

图 4-21 选择重传协议流程

选择重传协议发送方实现源代码如下。

```cpp
/*******************************************/
/*File:global.h*/
/*功能: 全局变量以及数据结构定义头文件*/
/*******************************************/
#ifndef _global_H_
#define _global_H_
#define _CRT_SECURE_NO_WARNINGS
#include<iostream>
#include<thread>
#include<mutex>
#include<condition_variable>
#include<windows.h>
#include<stdlib.h>
#include<pcap.h>
#include<string.h>
using namespace std;
//共享队列大小
const int kSizeOfQueue = 100 ;
//数据帧的最大长度, 因为能发送的最大长度是 1514 字节, 自动添加 4 字节 CRC 校验码
const int kSizeOfFrame = 1514;
const int kNumOfReadAndLoad = 1;         //read_and_load 线程的个数
const int kNumOfSendPacket = 1;          //send_packet 线程的个数
const int kNumOfAcceptAckEthernet = 1;   //accept_ack_ethernet 线程的个数
//数据帧首部的长度, 因为用 sizeof 会被对齐影响, 所以直接在此定义
const int kSizeOfHeader = 19;
const int kSizeOfCrc = 4;                //CRC 校验码的长度
const int kSizeOfWindows = 8;            //客户端发送窗口的大小
typedef struct ItemRepository{
```

```cpp
        u_int8_t item_queue[kSizeOfQueue][kSizeOfFrame];
        int seq_queue[16];        //对应队列中每个节点内容存放的数据帧序号
        int load_index;           //当前待加载的帧在queue中的对应下标
        int send_index;           //当前待发送的帧在queue中的对应下标
        int ack_index;            //等待收到的应答帧在queue中的对应下标
        int resend_index;
        mutex mtx;                //对队列访问的互斥量
        condition_variable cv_not_full;     //缓存队列未溢出信号
        condition_variable cv_not_empty;    //缓存队列不为空的信号
        condition_variable cv_got_ack;      //确认一帧发送完成并收到正确应答帧的信号
}IR;
void InitItemRepository(ItemRepository &ir);
void GenerateCrc32Table();
//kind =1:seq, kind=2:ack, 可直接输入枚举变量
int TurnToInt(u_int8_t *buffer, int kind);
#endif
#include"global.h"
#include"thread_send_ethernet.h"
#include"thread_accept_and_send.h"
#include"thread_read_and_load.h"
#include"fuc_load_frame.h"
#include"thread_time_clock.h"
int flag_read_end = 0;
int flag_load_end = 0;
int flag_send_end = 0;
IR gIR;
u_int32_t crc32_table[256];
extern thread thread_time_clock[kSizeOfQueue];
extern bool flag_arrive[kSizeOfQueue];
extern bool flag_over_time[kSizeOfQueue];
/*****************************************/
/*File: main.cpp*/
/*功能: main主函数执行入口实现文件*/
/*****************************************/
int main()
{
    InitItemRepository(gIR);
    GenerateCrc32Table();

    thread read_and_load[kNumOfReadAndLoad];
    for (int i = 0; i < kNumOfReadAndLoad; ++i){
        read_and_load[i] = thread(ReadAndLoad);
    }

    thread send_packet[kNumOfSendPacket];
    for (int i = 0; i < kNumOfSendPacket; ++i){
```

```cpp
        send_packet[i] = thread(SendEthernet);
    }

    thread accept_ack_ethernet[kNumOfAcceptAckEthernet];
    for (int i = 0; i < kNumOfAcceptAckEthernet; ++i){
        accept_ack_ethernet[i] = thread(AcceptAckEthernet);
    }

    for (int i = 0; i < kNumOfAcceptAckEthernet; ++i){
        if (accept_ack_ethernet[i].joinable())
        //可挂起当前线程(main)，直到accept_ack_ethernet[i]线程运行结束，再回来
        //运行当前线程(main)
            accept_ack_ethernet[i].join();    }
        //join()主要是用来挂起当前函数的，所以不需要考虑前面的join()会对后面的线程有影响
        //只会对当前线程(此处即main)造成阻塞
    for (int i = 0; i < kNumOfSendPacket; ++i){
        if (send_packet[i].joinable())
            send_packet[i].join();
    }

    for (int i = 0; i < kNumOfReadAndLoad; ++i){
        if (read_and_load[i].joinable())
            read_and_load[i].join();
    }

    for (int i = 0; i < gIR.send_index; ++i){
        if (thread_time_clock[i].joinable())
            thread_time_clock[i].join();
    }

    cout << "\n发送完成" << endl;
    system("pause");
    return 0;
}
void InitItemRepository(ItemRepository &ir){
    memset(ir.item_queue, 0, sizeof(ir.item_queue));
    ir.send_index = 0;
    ir.load_index = 0;
    ir.ack_index = 0;
    int i ;
    i = 0;
    while (i < kSizeOfQueue){
        flag_over_time[i] = false;
        flag_arrive[i] = false;
        ++i;
    }
```

```
}

void GenerateCrc32Table(){
    int i, j;
    u_int32_t crc;
    for (i = 0; i < 256; i++)
    {
        crc = i;
        for (j = 0; j < 8; j++)
        {
            if (crc & 1)
                crc = (crc >> 1) ^ 0xEDB88320;
            else
                crc >>= 1;
        }
        crc32_table[i] = crc;
    }
}
int TurnToInt(u_int8_t *buffer, int kind){//kind =1:seq, kind=2:ack
    if (1 == kind){
        return
            *buffer +
            (*(buffer+1)) * 256 +
            (*(buffer+2)) * 256 * 256 +
            (*(buffer+3)) * 256 * 256 * 256;
    }else
        if (2 == kind){
            return
                *buffer +
                (*(buffer + 1)) * 256 +
                (*(buffer + 2)) * 256 * 256 +
                (*(buffer + 3)) * 256 * 256 * 256;
        }
        else{
            cout << "Error:kind is error.\n" << endl;
        }
        return -1;
}
/*******************************************/
/*File: fuc_open_adapter.h*/
/*功能: 打开网络接口头文件*/
/*******************************************/
#ifndef _fuc_open_adapter_H_
#define _fuc_open_adapter_H_
pcap_t *OpenAdapter();
```

```cpp
#endif

/*****************************************/
/*File: fuc_open_adapter.cpp*/
/*功能: 打开网络接口实现文件*/
/*****************************************/
#include"global.h"
#include"fuc_open_adapter.h"
pcap_t *adhandle = OpenAdapter();
pcap_t *OpenAdapter(){
    pcap_if_t *alldevs;//pcap_if_t 和 pcap_if 几乎一样
    pcap_if_t *d;
    pcap_t *adhandle;
    int i = 0;
    char ErrBuf[PCAP_ERRBUF_SIZE];

    /*获取本地网络接口设备列表*/
    if (pcap_findalldevs_ex(PCAP_SRC_IF_STRING, NULL, &alldevs, ErrBuf) == -1){
        printf("\nError in pcap_findalldevs_ex : %s\n", ErrBuf);
        system("pause");
        exit(1);
    }
    /*打印本地网络接口设备列表*/
    for (d = alldevs; d != NULL; d = d->next){
        printf("%d.%s", ++i, d->name);
        if (d->description)
            printf("%s\n", d->description);
        else
            printf("\nNo descriptio available.\n");
    }
    if (i == 0){
        printf("\nNo interfaces found!Make sure WinPcap is installed.\n");
        pcap_freealldevs(alldevs);
        system("pause");
        exit(1);
    }

    /*选择本次通信网络接口*/
    int INum;
    while (1){
        printf("Enter the interface nuber (1-%d):", i);
        scanf("%d", &INum);
        if (INum<1 || INum>i){
            printf("\nInterface number out of range.\n");
        }
```

```c
        else break;
    }
    for (d = alldevs, i = 0; i < INum - 1; d = d->next, i++);

    /*打开选择的网络接口 adapter*/
    if ((adhandle = pcap_open_live(
        d->name,
        65536,
        1,
        1000,
        ErrBuf
        )) == NULL){
        printf("\nUnable to open the adapter. %s is not supported by Winpcap\n", d->name);
        pcap_freealldevs(alldevs);
        system("pause");
        exit(1);
    }
    else {
        pcap_freealldevs(alldevs);
        return adhandle;
    }
}


/*************************************/
/*File: fuc_load_frame.h*/
/*功能：构建数据帧头文件*/
/*************************************/
#ifndef _fuc_load_frame_H_
#define _fuc_load_frame_H_
//可接收的数据帧的源 MAC 地址
/const u_int8_t gAcceptedSourceMac[2][6] = { { 0xFF, 0xFF, 0xFF, 0xFF, 0xFF,
        0xFF }, { 0x3C, 0x77, 0xE6, 0x25, 0x91, 0x00 } };
//广播地址和本地 MAC 地址
const u_int8_t gClientMac[2][6] = { { 0xFF, 0xFF, 0xFF, 0xFF, 0xFF, 0xFF },
    { 0x3C, 0x77, 0xE6, 0x25, 0x91, 0x00 } };
//本地对协议类型采用小端存储，所以要将 0800 改为 0008，实际存储为低位-0800-高位，协议
//栈才能读出是 IPv4 协议类型
const u_int16_t gEthernetType = 0x0008;
//最大帧长定义
const int kSizeOfMaxFrame = 1514;       //14+3+1493+4=1514
//最小帧长定义
const int kSizeOfMinFrame = 64;         //14+46+4=64

enum DestinationMac{
    destinational_broadcast_mac = 1, destinational_client_mac
```

```cpp
};
enum SourceMac{
    source_broadcast_mac = 1, source_server_mac
};

struct EthernetHeader
{
    u_int8_t Preamble[8];              //前导码
    u_int8_t destination_mac[6];       //目标MAC地址
    u_int8_t source_mac[6];            //源MAC地址
    u_int16_t ethernet_type;           //协议类型
    u_int16_t identify;                //本程序构造的特殊数据帧标识
    u_int16_t len;      //整个数据帧的长度，因为有填充的缘故，所以将seq置于len之后
    u_int8_t seq;       //用来区别其他以太网数据帧的特殊标识字段
};

/*
 *功能：封装数据帧的数据字段
 *返回值：数据帧的数据字段字节个数，异常情况下返回-1
 */
void LoadFrame(u_int8_t *buffer, FILE *fp);

/*
 *功能：封装以太网数据帧的首部
 *返回值：执行成功返回1
 */
void LoadEthernetHeader(u_int8_t *buffer, u_int16_t len);
void LoadEthernetCrc(u_int8_t *buffer, int len);
void GenerateCrc32Table();
#endif

/*****************************************/
/*File: fuc_load_frame.cpp*/
/*功能：构建数据帧实现文件*/
/*****************************************/

#include"global.h"
#include"fuc_load_frame.h"
//0表示正确版本；1表示错误版本
int fault_version = 0;
extern int flag_read_end;
extern int flag_load_end;
extern u_int32_t crc32_table[256];
u_int8_t seq=0x00;
static int no = 0;
```

```cpp
void LoadFrame(u_int8_t *buffer, FILE *fp)
{
    memset(buffer, 0, kSizeOfFrame);

    int size_of_header;
    int size_of_data;
    int size_of_crc;

    size_of_header = kSizeOfHeader;
    size_of_crc = 4;

    /*读取数据帧的数据部分*/
    size_of_data = fread(
        buffer + size_of_header,
        1,
        kSizeOfMaxFrame-size_of_header-size_of_crc,
        fp);

    if (size_of_data < 0){           //先判断数据读取是否成功
        cout << "Error:Fread in LoadFrame is failed." << endl;
        return ;
    }

    if (0 == size_of_data){          //再判断数据是否已经读取完毕
        cout << "Success:Fread has fetched complately." << endl;
        flag_read_end = 1;
        return ;
    }
    //最后判断数据大小是否符合要求

    while (size_of_data < kSizeOfMinFrame -size_of_header-size_of_crc ){
        *(buffer + size_of_header + size_of_data) = 0;
        ++size_of_data;
    }

    /*构建帧首部*/
    LoadEthernetHeader(buffer,size_of_header+size_of_data+size_of_crc);
    //在相应位置填充帧首部

    /*构建CRC校验码*/
    if (fault_version == 1){
        if (no == 2){
            *(u_int32_t*)(buffer + size_of_header + size_of_data) = 0xffffffff;
        }
        else{
            //因为CRC校验是针对全部帧首部+数据部分的,所以需要在确定了seq之后再加上
```

```c
            LoadEthernetCrc(buffer, size_of_header + size_of_data);     }
        no++;
    }
    else{
        //因为 CRC 校验是对全部帧首部+数据部分的，所以需要在确定了 seq 之后再加上
        LoadEthernetCrc(buffer, size_of_header + size_of_data);
    }
    return ;
}

void LoadEthernetHeader(u_int8_t *buffer,u_int16_t len){
    struct EthernetHeader *header = (struct EthernetHeader*)buffer;
    /*目的 MAC 地址*/
    header->destination_mac[0] = gAcceptedSourceMac[1][0];
    header->destination_mac[1] = gAcceptedSourceMac[1][1];
    header->destination_mac[2] = gAcceptedSourceMac[1][2];
    header->destination_mac[3] = gAcceptedSourceMac[1][3];
    header->destination_mac[4] = gAcceptedSourceMac[1][4];
    header->destination_mac[5] = gAcceptedSourceMac[1][5];
    /*源 MAC 地址*/
    header->source_mac[0] = gClientMac[1][0];
    header->source_mac[1] = gClientMac[1][1];
    header->source_mac[2] = gClientMac[1][2];
    header->source_mac[3] = gClientMac[1][3];
    header->source_mac[4] = gClientMac[1][4];
    header->source_mac[5] = gClientMac[1][5];
    /*协议类型*/
    header->ethernet_type = gEthernetType;
    /*标识*/
    header->identify = gIdentify;
    header->seq = seq ;
    seq += 0x10;
    header->len = len;
    return ;
}

void LoadEthernetCrc(u_int8_t *buffer,int len){
    int i;
    u_int32_t crc;
    GenerateCrc32Table();
    crc = 0xffffffff;
    for (i = 0; i < len; i++)
    {
        crc = (crc >> 8) ^ crc32_table[(crc & 0xFF) ^ buffer[i]];
```

```c
        crc ^= 0xffffffff;
        *(u_int32_t*)(buffer + len) = crc;
}


/*******************************************/
/*File: thread_read_and_load.h*/
/*功能: 队列管理头文件*/
/*******************************************/
#ifndef _thread_read_and_load_H_
#define _thread_read_and_load_H_
void ReadAndLoad();
void LoadEthernetCrc(u_int8_t *buffer, int len);
void GenerateCrc32Table();
#endif


/*******************************************/
/*File: thread_read_and_load.cpp*/
/*功能: 队列管理实现文件*/
/*******************************************/
#include"global.h"
#include"thread_read_and_load.h"
#include"fuc_load_frame.h"

extern IR gIR;
extern int flag_read_end;
extern int flag_load_end;

void ReadAndLoad(){
    FILE *fp = fopen("data.txt", "r");
    if (NULL == fp){
        cout << "File opened error."<<endl;
        exit(1);
    }
    while (1){
        unique_lock<mutex> lock(gIR.mtx);           //上锁
        while (gIR.send_index == (gIR.load_index+1)%kSizeOfQueue){
            cout << "The shared queue is full, ReadAndLoad is waiting for an
                empty..."<<endl;
            (gIR.cv_not_full).wait(lock);           //等待收到缓存队列未溢出的信号
        }
        //将数据帧装载到该共享队列的 index 位置
        LoadFrame(gIR.item_queue[gIR.load_index], fp);
        struct EthernetHeader *ethernet_header = (struct EthernetHeader*)
            (gIR.item_queue[gIR.load_index]);
```

```cpp
        gIR.seq_queue[(ethernet_header->seq & 0xf0) / 16] = gIR.load_index;

        if (1 == flag_read_end){
            flag_load_end = 1;
        }
        else{
            ++(gIR.load_index);
            if (kSizeOfQueue == gIR.load_index){
                gIR.load_index = 0;
            }
        }

        (gIR.cv_not_empty).notify_all();        //发出缓存队列不为空的信号
        lock.unlock();                          //解锁

        if (1 == flag_load_end){
        cout << "\n*************All in thread_read_and_load is completed.";
        return;
        }
    }
}




/*******************************************/
/*File: thread_send_ethernet.h*/
/*功能: 打开网络接口头文件*/
/*******************************************/
#ifndef _thread_send_ethernet_H_
#define _thread_send_ethernet_H_
void SendEthernet();
void SendFrame(int index);
#endif
/*******************************************/
/*File: thread_send_ethernet.cpp*/
/*功能: 打开网络接口实现文件*/
/*******************************************/
#include"global.h"
#include"thread_send_ethernet.h"
#include"fuc_open_adapter.h"
#include"fuc_load_frame.h"
#include"thread_time_clock.h"
extern IR gIR;
extern u_int8_t received_ack;
extern u_int8_t sent_seq;
extern pcap_t *adhandle;
```

```cpp
extern int flag_load_end;
extern int flag_send_end;
extern int flag_ack_end;
extern thread thread_time_clock[kSizeOfQueue];
extern bool flag_arrive[kSizeOfQueue];
extern bool flag_over_time[kSizeOfQueue];

void SendEthernet()
{
    while (1){
    cout << "\n即将发送的帧序号: " << gIR.send_index << endl;
    unique_lock<mutex> lock(gIR.mtx);
    while (gIR.send_index == gIR.load_index){
    cout << "The shared queue is empty, SendEthernet is waiting for an item..."
            << endl;
    if (1 == flag_ack_end){
        flag_send_end = 1;
        cout << "\n***************All in thread_send_ethernet is
            completed." << endl;
        return;
    }

    (gIR.cv_not_empty).wait(lock);          //等待缓存队列不为空的信号
    if (true == flag_over_time[gIR.resend_index]){
        SendFrame(gIR.resend_index);
        }
    }
    //窗口加锁，等待ACK应答
    while (kSizeOfWindows - 1 <= (gIR.send_index - gIR.ack_index)){
    (gIR.cv_got_ack).wait(lock);
    }
    SendFrame(gIR.send_index);
    ++gIR.send_index;
    if (kSizeOfQueue == gIR.send_index)
        gIR.send_index = 0;
        (gIR.cv_not_full).notify_all();
        lock.unlock();

    }
}

void SendFrame(int index){
    struct EthernetHeader *ethernet_header = (struct EthernetHeader*)
        (gIR.item_queue[index]);
    //数据帧长度最大为1514字节：6+6+2+1500，不考虑CRC校验码
    if (0 != pcap_sendpacket(adhandle, (const u_char*)gIR.item_queue[index],
```

```
                    ethernet_header->len)){
cout << "Error:the errmsg of seq=" << index << " packet sended is:"
        << pcap_geterr(adhandle) << endl;
        flag_send_end = 1;    //直接置1表示结束
        //希望这里能补充发送失败时的重发机制
    }
    else {
        cout << "第 " << index << " 号窗口的数据发送成功,";
        //printf("%s\n", gIR.item_queue[index] + kSizeOfHeader);
        /*发送成功,开启相应的计时器*/
        if (thread_time_clock[index].joinable()){
            thread_time_clock[index].join();
        }
        flag_arrive[index] = false;
        flag_over_time[index] = false;
        thread_time_clock[index] = thread(TimeClock, index);
        cout <<"对应的计时器开启! " << endl;
        cout << "目前发送窗口左端窗口号: " << gIR.ack_index << " 发送窗口右端窗
            口号: " << gIR.send_index << endl;
    }
}


/****************************************/
/*File: thread_accept_and_send.h*/
/*功能: 接收ACK应答帧及首部检查头文件*/
/****************************************/
#ifndef _thread_accept_and_send_H_
#define _thread_accept_and_send_H_
void AcceptAckEthernet();
//CRC校验正确return 1, CRC校验错误return -1, 其他return 0
int HeaderCheck(const u_char* pkt_data);
bool CrcCheckOut(u_int8_t *buffer, int len);
#endif
/****************************************/
/*File: thread_accept_and_send.cpp*/
/*功能: 接收ACK应答帧及首部检查实现文件*/
/****************************************/
#include"global.h"
#include"thread_accept_and_send.h"
#include"fuc_open_adapter.h"
#include"thread_send_ethernet.h"
#include"fuc_load_frame.h"
#include"thread_time_clock.h"
```

```cpp
extern IR gIR;
u_int8_t received_ack;
extern pcap_t *adhandle;
extern int flag_send_end;
extern int flag_load_end;
int flag_ack_end;
struct pcap_pkthdr *header;
const u_char *pkt_data;
extern u_int32_t crc32_table[256];
extern bool flag_arrive[kSizeOfQueue];
extern bool flag_over_time[kSizeOfQueue];
extern thread thread_time_clock[kSizeOfQueue];
int header_check;
void AcceptAckEthernet(){
    int res = 0;
    while (1){
        unique_lock<mutex> lock(gIR.mtx);

        if ((res = pcap_next_ex(adhandle, &header, &pkt_data)) >= 0){
        //ahandle的超时时间到，但因为有计时器的存在，所以不考虑这个计时
        if (0 == res){
            continue;
            }

        if (-1 == res){
            cout << "Error(in thread_accept_and_send):Reading the packets:"
                << pcap_geterr(adhandle);
            }
        header_check = HeaderCheck(pkt_data);
        if (1 == header_check){        //CRC校验正确且是目标帧
        struct EthernetHeader* received_ethernet_header = (struct
            EthernetHeader*)pkt_data;
        received_ack = received_ethernet_header->seq;     //序号的校验
        int i, temp_index;
        i = 0;
        while ((gIR.ack_index + i) % kSizeOfQueue < gIR.send_index){
        temp_index = (gIR.ack_index + i) % kSizeOfQueue;
        struct EthernetHeader* temp_header = (struct EthernetHeader*)
            (gIR.item_queue[temp_index]);
        if (temp_header->seq+0x10 == received_ack){
            flag_arrive[temp_index] = true;
        if (thread_time_clock[temp_index].joinable()){
        thread_time_clock[temp_index].join();
            }
        cout << "收到 " << temp_index << " 号窗口发送的帧的应答帧";
        if (0 == i){
```

```cpp
                    gIR.ack_index++;
                    if (kSizeOfQueue == gIR.ack_index){

                    gIR.ack_index = 0;
                        }
                (gIR.cv_got_ack).notify_all();
                cout << "滑动窗口左端右移至:" << gIR.ack_index << endl;
                }
                else{
                    break;
                }
                cout << "目前发送窗口左端窗口号: " << gIR.ack_index << " 发送窗口右端窗
                    口号: " << gIR.send_index << endl;
                }
                else{
                        ++i;
                    }
                }
            }
            else{
                    continue;
            }
            if (gIR.ack_index == gIR.load_index){//检测是否收到了全部应答帧
                if (1 == flag_load_end){
                flag_ack_end = 1;
                cout << "\n********************All in thread_accept_and_send is
                        completed.";
                (gIR.cv_not_empty).notify_all();   //发出缓存队列不为空的信号
                return;
                }
                return;
                }
        }
    }
}
//CRC 校验正确 return 1, CRC 校验错误 return -1, 其他 return 0
int HeaderCheck(const u_char* pkt_data){
    struct EthernetHeader* ethernet_header = (struct EthernetHeader*)pkt_data;
    /*CRC 检验: 帧首部+数据部分校验*/
    if (false == CrcCheckOut((u_int8_t*)pkt_data, ethernet_header->len -
            kSizeOfCrc))return -1;
    /*协议类型校验*/
    if (ethernet_header->ethernet_type != gEthernetType){
        return 0;
    }
    /*MAC 地址校验*/
```

```c
int i = 0, j = 0;
//判断是否为发送到客户端对应的MAC地址
int destination_flag = 0;
for (i = 0; i < 2; ++i){
    for (j = 0; j < 6; ++j){
        if (ethernet_header->destination_mac[j] != gClientMac[i][j])
           break;
    }
    if (6 == j){
        destination_flag = i + 1;//flag=1:broadcast_mac  flag=2:client_mac
        break;
    }
}
if (0 == destination_flag){
    return 0;
}
if (destinational_broadcast_mac == destination_flag){
    return 0;
}
//目的MAC地址为发送到客户端地址的以太网数据帧
if (destinational_client_mac == destination_flag){
//判断是否为可接收的源MAC地址
    int source_flag = 0;
    for (i = 0; i < 2; ++i){
        for (j = 0; j < 6; ++j){
            if (ethernet_header->source_mac[j] != gAcceptedSourceMac[i][j])
            break;
        }
        if (6 == j){
            source_flag = i + 1;
            break;
        }
    }
    if (0 == source_flag){
        return 0;
    }
    if (source_broadcast_mac == source_flag){
        return 0;
    }
    //判断是否为可接收的服务器端的地址
    if (source_server_mac == source_flag){
        return 1;
    }
}
return 0;
}
```

```cpp
bool CrcCheckOut(u_int8_t *buffer, int len){
    int i;
    u_int32_t crc;

    crc = 0xffffffff;
    for (i = 0; i < len; i++)
    {
        crc = (crc >> 8) ^ crc32_table[(crc & 0xFF) ^ buffer[i]];
    }
    crc ^= 0xffffffff;

    if (*(u_int32_t*)(buffer + len) == crc)
        return true;

    return false;
}




/*****************************************/
/*File: thread_time_clock.h*/
/*功能：打开网络接口头文件*/
/*****************************************/
#ifndef _thread_time_clock_H_
#define _thread_time_clock_H_
void TimeClock(int index);
#endif
/*****************************************/
/*File: fuc_open_adapter.cpp*/
/*功能：打开网络接口实现文件*/
/*****************************************/

#include"global.h"
#include"thread_time_clock.h"
#include"thread_send_ethernet.h"
bool flag_arrive[kSizeOfQueue];
bool flag_over_time[kSizeOfQueue];
thread thread_time_clock[kSizeOfQueue];
extern IR gIR;
void TimeClock(int index){
    //超时时间设为5s
    int seconds = 5;
    while (seconds > 0){
        if (true == flag_arrive[index])return;
        Sleep(1000);
```

```
            seconds--;
        }

        if (0 == seconds){
            //激活重发
            flag_over_time[index] = true;
            cout << "第 " << index <<" 号窗口对应的帧超时,即将重传! "<< endl;
            gIR.resend_index = index;
            gIR.cv_not_empty.notify_all();
            //SendFrame(index);
            return;
        }
        return;
    }
```

选择重发协议接收端实现源代码如下。

```
/******************************************/
/*File: global.h*/
/*功能: 全局变量、数据结构头文件*/
/******************************************/
#ifndef _global_H_
#define _global_H_
#define _CRT_SECURE_NO_WARNINGS
#include<iostream>
#include<mutex>
#include<pcap.h>
#include<thread>
#include<string.h>
#include<condition_variable>
using namespace std;
const int kSizeOfQueue = 100;
const int kSizeOfFrame = 1514;
const int kNumOfExtractAndWrite = 1;
const int kNumOfReceiveEthernet = 1;
const int kPcapErrbufSize = 256;
const int kSizeOfHeader = 19;
const int kSizeOfCrc = 4;
const int kSizeOfWindows = 8;
const int kNumOfSendAckEthernet = 1;
typedef struct ItemRepository{
    u_int8_t item_queue[kSizeOfQueue][kSizeOfFrame];
    int receive_index;
    int extract_index;
    int ack_index;
    mutex mtx;
```

```cpp
    condition_variable cv_not_full;            //缓存队列未溢出信号
    condition_variable cv_not_empty;           //缓存队列不为空的信号
    condition_variable cv_extract;
    condition_variable cv_got_seq;             //收到帧信号
}IR;
enum SeqOrAck{
    seq = 1, ack
};

void InitItemRepository(ItemRepository &ir);

void GenerateCrc32Table();
//kind =1:seq，kind=2:ack，可直接输入枚举变量
int TurnToInt(u_int8_t *buffer, int kind);
#endif


/**************************************/
/*File: main.cpp*/
/*功能：main 主函数执行入口实现文件*/
/**************************************/
#include"global.h"
#include"thread_receive_load_ack.h"
#include"thread_extract_and_write.h"
#include"fuc_send_ack_ethernet.h"

u_int8_t received_seq=0;
u_int8_t sent_ack=0;
IR gIR;
u_int32_t crc32_table[256];
int main(){
    InitItemRepository(gIR);
    GenerateCrc32Table();

    thread receive_ethernet[kNumOfReceiveEthernet];
    for (int i = 0; i < kNumOfReceiveEthernet; ++i){
        receive_ethernet[i] = thread(ReceiveEthernet);
    }
    thread extract_and_write[kNumOfExtractAndWrite];
    for (int i = 0; i < kNumOfExtractAndWrite; ++i){
        extract_and_write[i] = thread(ExtractAndWrite);
    }
    for (int i = 0; i < kNumOfReceiveEthernet; ++i){
        receive_ethernet[i].join();
    }
    for (int i = 0; i < kNumOfExtractAndWrite; ++i){
```

```
            extract_and_write[i].join();
    }
    while (1);
    return 0;
}

void InitItemRepository(ItemRepository &ir){
    ir.extract_index = 0;
    ir.receive_index = 0;
    ir.ack_index = 0;
}

void GenerateCrc32Table(){
    int i, j;

    u_int32_t crc;
    for (i = 0; i < 256; i++)
    {
        crc = i;
        for (j = 0; j < 8; j++)
        {
            if (crc & 1)
                crc = (crc >> 1) ^ 0xEDB88320;
            else
                crc >>= 1;
        }
        crc32_table[i] = crc;
    }
}
int TurnToInt(u_int8_t *buffer, int kind){//kind =1:seq, kind=2:ack
    if (1 == kind){
        return
            *buffer +
            (*(buffer + 1)) * 256 +
            (*(buffer + 2)) * 256 * 256 +
            (*(buffer + 3)) * 256 * 256 * 256;
    }
    else
        if (2 == kind){
            return
                *buffer +
                (*(buffer + 1)) * 256 +
                (*(buffer + 2)) * 256 * 256 +
                (*(buffer + 3)) * 256 * 256 * 256;
        }
        else{
```

```cpp
            cout << "Error:kind is error.\n" << endl;
        }
        return -1;
}


/*****************************************/
/*File: fuc_open_adapter.h*/
/*功能: 打开网络接口头文件*/
/*****************************************/
#ifndef _fuc_open_adapter_H_
#define _fuc_open_adapter_H_
pcap_t *OpenAdapter();
#endif


/*****************************************/
/*File: fuc_open_adapter.cpp*/
/*功能: 打开网络接口实现文件*/
/*****************************************/
#include"global.h"
#include"fuc_open_adapter.h"

pcap_t *adhandle = OpenAdapter();
pcap_t *OpenAdapter(){
    pcap_if_t *alldevs; //pcap_if_t 和 pcap_if 几乎一样
    pcap_if_t *d;
    pcap_t *adhandle;
    int i = 0;
    char ErrBuf[PCAP_ERRBUF_SIZE];

    /*获取本地网络接口设备列表*/
    if (pcap_findalldevs_ex(PCAP_SRC_IF_STRING, NULL, &alldevs, ErrBuf) == -1){
        printf("\nError in pcap_findalldevs_ex : %s\n", ErrBuf);
        system("pause");
        exit(1);
    }

    /*打印本地网络接口设备列表*/
    for (d = alldevs; d != NULL; d = d->next){
        printf("%d.%s", ++i, d->name);
        if (d->description)
            printf("%s\n", d->description);
        else
            printf("\nNo descriptio available.\n");
    }
    if (i == 0){
```

```c
        printf("\nNo interfaces found!Make sure WinPcap is installed.\n");
        pcap_freealldevs(alldevs);
        system("pause");
        exit(1);
    }

    /*选择本次通信网络接口*/
    int INum;
    while (1){
        printf("Enter the interface nuber (1-%d):", i);
        scanf("%d", &INum);
        if (INum<1 || INum>i){
            printf("\nInterface number out of range.\n");
        }
        else break;
    }
    for (d = alldevs, i = 0; i < INum - 1; d = d->next, i++);

    /*打开选择的网络接口 adapter*/
    if ((adhandle = pcap_open_live(
        d->name,
        65536,
        1,
        1000,
        ErrBuf
        )) == NULL){
    printf("\nUnable to open the adapter. %s is not supported by Winpcap\n",
            d->name);
        pcap_freealldevs(alldevs);
        system("pause");
        exit(1);
    }
    else {
        pcap_freealldevs(alldevs);
        return adhandle;
    }
}

/*******************************************/
/*File: thread_receive_load_ack.h*/
/*功能: 数据帧接收及首部检测头文件*/
/*******************************************/
#ifndef _thread_receive_load_ack_H_
#define _thread_receive_load_ack_H_
void ReceiveEthernet();
//CRC 校验正确 return 1, CRC 校验错误 return -1, 其他 return 0
```

```cpp
int HeaderCheck(const u_char* pkt_data);
bool CrcCheckOut(u_int8_t *buffer, int len);
#endif

/*****************************************/
/*File: thread_receive_load_ack.cpp*/
/*功能：数据帧接收及首部检测实现文件*/
/*****************************************/
#include"global.h"
#include"thread_receive_load_ack.h"
#include"fuc_send_ack_ethernet.h"
#include"fuc_open_adapter.h"
#include"fuc_analyze_frame.h"

extern u_int8_t received_seq;
extern u_int8_t sent_ack;
extern IR gIR;
extern pcap_t *adhandle;
extern u_int32_t crc32_table[256];
int res;
int source_index = 0;
struct pcap_pkthdr *header;
const u_char *pkt_data;
char errbuf[kPcapErrbufSize];
int header_check;
void ReceiveEthernet(){
    int res;
    while (1){
    //超时相关控制
    if ((res = pcap_next_ex(adhandle, &header, &pkt_data)) >= 0){
    if (0 == res)
       continue;
    header_check = HeaderCheck(pkt_data);
    //校验无差错且源 MAC 地址和目的 MAC 地址相符
    if (1 == header_check){
    struct EthernetHeader* ethernet_header = (struct EthernetHeader*)pkt_data;
    received_seq = ethernet_header->seq;
    unique_lock<mutex> lock(gIR.mtx);         //互斥锁
    while ((gIR.receive_index + 1) % kSizeOfQueue == gIR.extract_index){
    cout << "The shared queue is full, thread_receive_load_ack is waiting
           for an empty...\n";
    gIR.cv_not_full.wait(lock);
    }
    //因窗口大小问题被锁住，等待 extract 信号
    while (kSizeOfWindows-1 <= (gIR.receive_index - gIR.extract_index)){
       gIR.cv_extract.wait(lock);
```

```cpp
            }
            //发送应答帧
            SendAckEthernet(received_seq+0x10, source_index);
            int k = 0;
            //写入共享队列，包括帧首部、数据部分和 CRC 校验码
            for (u_int8_t *p = (u_int8_t*)pkt_data; p != (u_int8_t*)(pkt_data +
                ethernet_header->len); p++)
            gIR.item_queue[gIR.receive_index][k++] = *p;
            printf("Loaded data in thread_receive_load_ack:\n%d:\n%s\n",
                gIR.receive_index, gIR.item_queue[gIR.receive_index] +
                kSizeOfHeader);
            ++gIR.receive_index;
            if (kSizeOfQueue == gIR.receive_index){
                gIR.receive_index = 0;
            }

            cout << "\n滑动窗口右端右移至:" << gIR.receive_index << endl;
            cout << "目前发送窗口左端窗口号: " << gIR.extract_index << " 发送窗口右端
                    窗口号: " << gIR.receive_index << endl;
            gIR.cv_not_empty.notify_all();
            lock.unlock();
        }
        else if (-1 == header_check){
            //若 CRC 校验的结果显示数据传输出错，则不可对 sent_ack 进行重发，直接等对方超时重传
            continue;
        }
        if (-1 == res){
            cout << "Error(in thread_accept_and_send):Reading the packets:"
                << pcap_geterr(adhandle);
            }
        }
    }
}
//CRC 校验正确 return 1, CRC 校验错误 return -1, 其他 return 0
int HeaderCheck(const u_char* pkt_data){
    struct EthernetHeader* ethernet_header = (struct EthernetHeader*)pkt_data;
    //帧首部+数据部分校验
    if (false == CrcCheckOut((u_int8_t*)pkt_data, ethernet_header->len -
            kSizeOfCrc)) return -1;

    /*协议类型校验*/
    if (ethernet_header->ethernet_type != gEthernetType){
        return 0;
    }
    /*MAC 地址校验*/
    //source_index 用来指示对可接收范围内哪个客户端 MAC 地址进行反馈
```

```c
    int i = 0, j = 0;
    //判断是否为发送到服务器端对应的MAC地址
    int destination_flag = 0;
    for (i = 0; i < 2; ++i){
    for (j = 0; j < 6; ++j){
        if (ethernet_header->destination_mac[j] != gServerMac[i][j])
            break;
    }
    if (6 == j){
        destination_flag = i + 1;//flag=1:broadcast_mac   flag=2:server_mac
        break;
        }
    }
    if (0 == destination_flag){
        return 0;
    }
    if (destinational_broadcast_mac == destination_flag){
        return 0;
    }
    //目的MAC地址为发送到服务器端地址的以太网数据帧
    if (destinational_server_mac == destination_flag){
    /*判断是否为可接收的地址*/
    int source_flag = 0;
    for (i = 0; i < 2; ++i){
    for (j = 0; j < 6; ++j){
    if (ethernet_header->source_mac[j] != gAcceptedSourceMac[i][j])
        break;
        }
    if (6 == j){
        source_flag = i + 1;
        source_index = i;
        break;
        }
        }
        if (0 == source_flag){
            return 0;
        }
        if (source_broadcast_mac == source_flag){
            return 0;
        }
        if (source_client_mac == source_flag){
            return 1;
        }
    }
    return 0;
}
```

```cpp
bool CrcCheckOut(u_int8_t *buffer, int len){
    int i;
    u_int32_t crc;
    crc = 0xffffffff;

    for (i = 0; i < len; i++)
    {
        crc = (crc >> 8) ^ crc32_table[(crc & 0xFF) ^ buffer[i]];
    }
    crc ^= 0xffffffff;
    if (*(u_int32_t*)(buffer + len) == crc)
        return true;
    return false;
}

/***************************************/
/*Fle: thread_receive_load_ack.h*/
/*功能：接收数据帧并进行首部检查头文件*/
/***************************************/
#ifndef _thread_receive_load_ack_H_
#define _thread_receive_load_ack_H_
void ReceiveEthernet();
//CRC校验正确return 1, CRC校验错误return -1, 其他return 0
int HeaderCheck(const u_char* pkt_data);
bool CrcCheckOut(u_int8_t *buffer, int len);
#endif
/***************************************/
/*File: thread_receive_load_ack.cpp*/
/*功能：接收数据帧并进行首部检查实现文件*/
/***************************************/
#include"global.h"
#include"thread_receive_load_ack.h"
#include"fuc_send_ack_ethernet.h"
#include"fuc_open_adapter.h"
#include"fuc_analyze_frame.h"

extern u_int8_t received_seq;
extern u_int8_t sent_ack;
extern IR gIR;
extern pcap_t *adhandle;
extern u_int32_t crc32_table[256];
int res;
int source_index = 0;
struct pcap_pkthdr *header;
const u_char *pkt_data;
```

```
char errbuf[kPcapErrbufSize];
int header_check;

void ReceiveEthernet(){
    int res;
    while (1){
    //超时相关控制
    if ((res = pcap_next_ex(adhandle, &header, &pkt_data)) >= 0){
    if (0 == res)
        continue;
    header_check = HeaderCheck(pkt_data);
    //校验无差错且源 MAC 地址和目的 MAC 地址相符
    if (1 == header_check){
    struct EthernetHeader* ethernet_header = (struct EthernetHeader*)pkt_data;
    received_seq = ethernet_header->seq;

    unique_lock<mutex> lock(gIR.mtx);        //互斥锁
    while ((gIR.receive_index + 1) % kSizeOfQueue == gIR.extract_index){
        cout << "The shared queue is full, thread_receive_load_ack is waiting
                for an empty...\n";
        gIR.cv_not_full.wait(lock);
    }
    //因窗口大小问题被锁住,等待 extract 信号
    while (kSizeOfWindows-1 <= (gIR.receive_index - gIR.extract_index)){
            gIR.cv_extract.wait(lock);
    }
    //发送应答帧
    SendAckEthernet(received_seq+0x10, source_index);
    int k = 0;
    //写入共享队列,包括帧首部、数据部分和 CRC 校验码
    for (u_int8_t *p = (u_int8_t*)pkt_data; p != (u_int8_t*)(pkt_data +
            ethernet_header->len); p++)
    gIR.item_queue[gIR.receive_index][k++] = *p;
    printf("Loaded data in thread_receive_load_ack:\n%d:\n%s\n",
        gIR.receive_index, gIR.item_queue[gIR.receive_index] + kSizeOfHeader);
    ++gIR.receive_index;
    if (kSizeOfQueue == gIR.receive_index){
        gIR.receive_index = 0;
    }
    cout << "\n滑动窗口右端右移至:" << gIR.receive_index << endl;
    cout << "目前发送窗口左端窗口号: " << gIR.extract_index << " 发送窗口右端
            窗口号: " << gIR.receive_index << endl;
    gIR.cv_not_empty.notify_all();
    lock.unlock();
    }
    else if (-1 == header_check){
```

```
        //若CRC校验的结果显示数据传输出错，则不要求对sent_ack进行重发，直接等对方超时重传
        continue;
    }
    if (-1 == res){
    cout << "Error(in thread_accept_and_send):Reading the packets:"
     << pcap_geterr(adhandle);
    }
  }
 }
}

    //CRC校验正确return 1，CRC校验错误return -1，其他return 0
    int HeaderCheck(const u_char* pkt_data){
    struct EthernetHeader* ethernet_header = (struct EthernetHeader*)pkt_data;
    //帧首部+数据部分校验
    if (false == CrcCheckOut((u_int8_t*)pkt_data, ethernet_header->len -
        kSizeOfCrc))
        return -1;

    /*协议类型校验*/
    if (ethernet_header->ethernet_type != gEthernetType){
        return 0;
    }
    /*MAC地址校验*/
    //source_index用来指示对可接收范围内哪个客户端MAC地址进行反馈
    int i = 0, j = 0;
    //判断是否为发送到服务器端的MAC地址
    int destination_flag = 0
    for (i = 0; i < 2; ++i){
       for (j = 0; j < 6; ++j){
    if (ethernet_header->destination_mac[j] != gServerMac[i][j])
         break;
    }
    if (6 == j){
        destination_flag = i + 1;//flag=1:broadcast_mac   flag=2:server_mac
        break;
      }
    }
    if (0 == destination_flag){
        return 0;
    }
    if (destinational_broadcast_mac == destination_flag){
        return 0;
    }
    //目的MAC地址为发送到服务器端地址的以太网数据帧
    if (destinational_server_mac == destination_flag){
    /*判断是否为可接收的地址*/
```

```c
    int source_flag = 0;
    for (i = 0; i < 2; ++i){
        for (j = 0; j < 6; ++j){
            if (ethernet_header->source_mac[j] != gAcceptedSourceMac[i][j])
                break;
        }
        if (6 == j){
            source_flag = i + 1;
            source_index = i;
            break;
        }
    }
    if (0 == source_flag){
        return 0;
    }
    if (source_broadcast_mac == source_flag){
        return 0;
    }
    if (source_client_mac == source_flag){
        return 1;
    }
}
return 0;
}

bool CrcCheckOut(u_int8_t *buffer, int len){
    int i;
    u_int32_t crc;
    crc = 0xffffffff;

    for (i = 0; i < len; i++)
    {
        crc = (crc >> 8) ^ crc32_table[(crc & 0xFF) ^ buffer[i]];
    }
    crc ^= 0xffffffff;
    if (*(u_int32_t*)(buffer + len) == crc)
        return true;
    return false;
}

/*******************************/
/*File: fuc_send_ack_ethernet.h*/
/*功能：构建数据帧首部及发送ACK应答帧头文件*/
/*******************************/
#ifndef _fuc_send_ack_ethernet_H_
#define _fuc_send_ack_ethernet_H_
```

```c
int SendAckEthernet(u_int8_t ack, int source_index);

/*
功能: 封装以太网数据帧的首部
返回值: 执行成功返回1
*/
void LoadEthernetHeader(u_int8_t *buffer ,u_int8_t ack,int source_index);
void LoadEthernetCrc(u_int8_t *buffer, int len);
#endif

/*******************************************/
/*File: fuc_send_ack_ethernet.cpp*/
/*功能: 构建数据帧首部及发送ACK应答帧实现文件*/
/*******************************************/
#include"global.h"
#include"fuc_send_ack_ethernet.h"
#include"fuc_analyze_frame.h"
#include"fuc_open_adapter.h"

extern pcap_t *adhandle;
extern u_int8_t sent_ack;
extern u_int8_t sent_seq;
extern u_int32_t crc32_table[256];

int SendAckEthernet(u_int8_t ack, int source_index){
    u_int8_t ack_frame[kSizeOfHeader+kSizeOfCrc];

    /*封装ACK应答帧首部*/
    LoadEthernetHeader(ack_frame,ack,source_index);
    sent_ack = ack;

    /*封装ACK应答帧CRC校验码字段*/
    LoadEthernetCrc(ack_frame, kSizeOfHeader);

    if (0 != pcap_sendpacket(adhandle, (const u_char*)ack_frame, kSizeOfHeader+
        kSizeOfCrc)){
        cout << "Warning:Ack sent error.\n";
        return -1;
    }
    else{
        cout << "Success:Ack sent successfully.\n";
        return 1;
    }
```

```c
}

void LoadEthernetHeader(u_int8_t *buffer,u_int8_t ack, int source_index){
    struct EthernetHeader *header = (struct EthernetHeader*)buffer;

    /*目的 MAC 地址*/
    header->destination_mac[0] = gAcceptedSourceMac[source_index][0];
    header->destination_mac[1] = gAcceptedSourceMac[source_index][1];
    header->destination_mac[2] = gAcceptedSourceMac[source_index][2];
    header->destination_mac[3] = gAcceptedSourceMac[source_index][3];
    header->destination_mac[4] = gAcceptedSourceMac[source_index][4];
    header->destination_mac[5] = gAcceptedSourceMac[source_index][5];

    /*源 MAC 地址*/
    header->source_mac[0] = gServerMac[source_index][0];
    header->source_mac[1] = gServerMac[source_index][1];
    header->source_mac[2] = gServerMac[source_index][2];
    header->source_mac[3] = gServerMac[source_index][3];
    header->source_mac[4] = gServerMac[source_index][4];
    header->source_mac[5] = gServerMac[source_index][5];

    /*协议类型*/
    header->ethernet_type = gEthernetType;

    header->len = kSizeOfHeader+kSizeOfCrc;

    header->identify = gIdentify;

    header->seq = ack;
}


void LoadEthernetCrc(u_int8_t *buffer, int len){
    int i;
    GenerateCrc32Table();
    u_int32_t crc;
    crc = 0xffffffff;
    for (i = 0; i < len; i++)
    {
        crc = (crc >> 8) ^ crc32_table[(crc & 0xFF) ^ buffer[i]];
    }
    crc ^= 0xffffffff;
    *(u_int32_t*)(buffer + len) = crc;
}

/*****************************************/
```

```cpp
/*File: thread_extract_and_write.h*/
/*功能：获取数据帧数据部分并将其写入文件头文件*/
/*****************************************/
#ifndef _thread_extract_and_write_H_
#define _thread_extract_and_write_H_
void ExtractAndWrite();
void WriteData(u_int8_t *buffer, int len,FILE* fp);
#endif
/*****************************************/
/*File: thread_extract_and_write.cpp*/
/*功能：获取数据帧数据部分并将其写入文件实现文件*/
/*****************************************/
#include"global.h"
#include"thread_extract_and_write.h"
#include"fuc_analyze_frame.h"
#include"fuc_send_ack_ethernet.h"

extern IR gIR;
u_int8_t extracted_seq = 0x00;

void ExtractAndWrite(){
    FILE *fp;

    /*打开文件*/
    fp = fopen("data_get.txt", "w");
    if (NULL == fp){
        cout << "File opend error.\n";
        return;
    }
    fclose(fp);
    while (1){
    unique_lock<mutex> lock(gIR.mtx);
    while (gIR.extract_index  == gIR.receive_index){
    cout << "The shared queue is empty, ExtractAndWrite is waiting for an itme...\n";
    //等待收到缓存队列不为空的信号
    (gIR.cv_not_empty).wait(lock);
    }
    int i;
    int temp_index;
    i = 0;
    while ((gIR.extract_index + i)%kSizeOfQueue < gIR.receive_index){
    temp_index = (gIR.extract_index + i) % kSizeOfQueue;
    struct EthernetHeader *ethernet_header = (struct EthernetHeader*)
        (gIR.item_queue[temp_index]);
    if (ethernet_header->seq == 0x00){
        extracted_seq = 0x00;
```

```
        }
        if (ethernet_header->seq == extracted_seq){
            cout << "第 " << temp_index<<" 号窗口的帧将被提交" << endl;
            WriteData(gIR.item_queue[temp_index] + kSizeOfHeader,ethernet_
                header->len - kSizeOfHeader - kSizeOfCrc, fp);
            extracted_seq = ethernet_header->seq+0x10;
            if (0 == i){
                ++gIR.extract_index;
            if (kSizeOfQueue == gIR.extract_index){
                gIR.extract_index = 0;
            }
             cout << "滑动窗口左端右移至:" << gIR.extract_index << endl;
            cout << "目前发送窗口左端窗口号: " << gIR.extract_index << " 发送窗口右端
                窗口号: " << gIR.receive_index << endl;
            (gIR.cv_extract).notify_all();
            (gIR.cv_not_full).notify_all();
            }
            i = 0;
            }
            else{
                ++i;
            }
        
        }
        
        cout << "in thread_extract_and_write:\nheader->len:" << header->len << endl;
        cout << "header->len-kSizeOfHeader-kSizeOfCrc:" << header->len -
            kSizeOfHeader - kSizeOfCrc << endl;
        
        (gIR.cv_not_full).notify_all();
        lock.unlock();
    }
}

void WriteData(u_int8_t *buffer, int len,FILE* fp){
    //追加方式写入 data_get.txt 文件
    fp = fopen("data_get.txt", "a");
    if (NULL == fp){
        cout << "File opend error.\n";
        return;
    }
    //从缓存区获取数据并写入 data-get.txt 文件
    fwrite(buffer, sizeof(u_int8_t), len, fp);
    fclose(fp);
    return;
}
```

第 5 章 网络层 ARP 协议分析与实践

5.1 概 述

网络层对互联网提供的网络层服务进行了抽象，目的是屏蔽下层复杂的实现细节，协议栈设计者可在网络层基础上提供端到端的数据传输服务，使网络层能够使用统一的、抽象的 IP 地址、路由协议、路由选择方法等实现主机和主机或主机和路由器之间的通信。在逻辑的网络层可以看到 IP 分组流，在数据链路层可以看到不同的数据帧，在实际物理链路上只能看到物理信号；数据链路层传输的数据帧需要在其首部封装目的 MAC 地址和源 MAC 地址，而且在二层及三层网络设备连接的不同物理网段，这两个地址发生了变化。因此，主机、二层及三层网络设备在发送网络接口封装数据帧时，需要确定目的 MAC 地址和源 MAC 地址的具体内容。

为了使 TCP/IP 协议与具体的 MAC 地址无关，通过网络层将 MAC 地址隐藏起来，在网络层统一使用 IP 地址进行网络交换通信。在同一网段内，由于所有网络接口的网络地址相同，不存在网络间的路由，因此必须借助目的 MAC 地址实现网段内不同二层网络设备间的转发。为此，必须在网络层实现一种解决 IP 地址与 MAC 地址之间的映射服务。实现该服务的协议称为地址解析协议（Address Resolution Protocol，ARP），如图 5-1 所示。同样，在网络层也需要提供一种从 MAC 地址到 IP 地址的映射服务，实现该服务的协议称为逆向地址解析协议（Reverse Address Resolution Protocol，RARP），如图 5-2 所示。

图 5-1　ARP 协议功能图　　　　图 5-2　RARP 协议功能图

如果一个主机（或其他网络设备）初始化后只有自己的 MAC 地址而没有 IP 地址，则该主机可以通过 RARP 协议发送广播式请求报文向 RARP 服务器请求获得自己的 IP 地址，RARP 服务器负责对该请求做出应答。所以，无 IP 地址的主机可以通过 RARP 协议来获取属于自己的 IP 地址。RARP 协议主要通过无盘工作站来获取网络设备的 IP 地址。RARP 报文格式与 ARP 报文格式相同。如果发送方以广播方式发送 RARP 请求报文，则在发送方 MAC 地址字段和接收方 MAC 地址字段都填入本机 MAC 地址。接收方接收到该请求报文后，根据一定的 IP 分配策略，自动给发送方回送一个 RARP 响应报文，发送方可从该报文的接收方 IP 地址字段中获得自己的 IP 地址。目前，在 IPv4 版本下，RARP 协议可以简单认为已经被应用层的 DHCP 协议提供的网络服务代替。

5.2 ARP 协议工作原理

不管网络层使用什么协议，在实际数据链路上传送数据帧时，同一网段内的二层转发必须依据网络接口的 MAC 地址的。每一个主机（或其他网络设备）都设有一个 ARP 高速缓存，用以保存在本地局域网范围内所能知道的所有主机和交换机的 IP 地址到 MAC 地址的映射关系，每一个映射关系作为一条记录都保存在内存中并具有一定生存期。当然，每个路由器也设有一个 ARP 高速缓存，用于保存其知道的所有 IP 地址到 MAC 地址的映射关系，并且每一个映射关系作为一条记录都具有生存期。当主机 A 欲向本地局域网上的主机 B 发送 IP 分组时，主机 A 先在其 ARP 高速缓存中查看是否存在主机 B 的 IP 地址到 MAC 地址的映射关系；如果存在且在生存期内，则可查出其对应的 MAC 地址，再将此 MAC 地址填写到数据帧的目的 MAC 地址字段，然后通过局域网将该数据帧发往主机 B；否则，启动 ARP 协议，获得主机 B 的 IP 地址对应的 MAC 地址。

为了减少网络上的通信量，主机 A 在发送其 ARP 请求报文时，该报文在数据链路层以二层广播帧的形式发送，并将发送方 IP 地址到 MAC 地址的映射关系写入 ARP 请求分组。当主机 B 收到主机 A 的 ARP 请求分组时，将主机 A 的这一地址映射关系写入主机 B 的 ARP 高速缓存中（具有生存期），当主机 B 再向主机 A 发送报文时就更方便了。只要该映射关系在生存期内，主机 B 可以直接从本地缓存中获取主机 A 的 MAC 地址，IP 协议也不用调用 ARP 协议，防止接收方为解析发送方 MAC 地址而发送 ARP 请求，造成网络资源浪费。

利用 arp –a 命令可以查看主机的 ARP 缓存内容。

由此可见，为了提高 ARP 协议工作效率，方法一是在 ARP 请求报文中放入源主机地址映射关系（IP 地址+MAC 地址）；方法二是在新主机加入网络时，主动广播自己的地址映射关系，以避免其他主机对它的地址进行解析。ARP 协议工作效率的提高会带来网络安全问题，如断网、网络时好时坏（某台主机或整个网络）、中间人攻击等，目前主要采用 MAC 地址与端口及 IP 地址绑定技术来解决。

5.2.1 ARP 协议语法

当一个主机向另一个主机发送报文时，只有知道与对方 IP 地址对应的 MAC 地址才能在网络上进行传输。这种地址解析服务是由 ARP 协议提供的。应当注意的是，ARP 协议只用于解析目的方主机（或网关）的 MAC 地址，而不是解析发送方主机的 MAC 地址。

ARP 协议采用动态联编（dynamic binding）方式来解析对方的 MAC 地址。具体来说，发送方主机向网络中广播一个 ARP 请求报文，ARP 请求报文在数据链路层被封装为二层广播帧，ARP 请求报文中包含有目的方主机的 IP 地址，目的是请求与该 IP 地址相符的目的方主机 MAC 地址。网络上所有的主机在网络层都能接收到这个 ARP 请求报文，但只有 IP 地址与 ARP 请求报文中目的方主机的 IP 地址相符合的主机才能发送一个 ARP 响应报文

（二层单播帧），并报告给发送方主机，目的方主机的 MAC 地址。这样，发送方主机就得到了目的方主机的 MAC 地址了。ARP 报文格式（语法）如图 5-3 所示。

在 ARP 报文格式中，具体内容如下。

网络类型：2 个字节，表示发送方主机的网络类型，其中，"1"代表以太网。

协议类型：2 个字节，表示发送方使用 ARP 协议获取 MAC 地址的高层协议类型，其中，"0x0800"代表 IP 协议。由于 ARP 协议是配合 IP 协议来提供地址解析服务的，因此在网络层一般是 IP 协议调用 ARP 协议的，所以协议类型一般为 IP 协议类型。

| 网络类型 |
| 协议类型 |
| MAC地址长度 |
| IP地址长度 |
| 操作 |
| 发送方MAC地址 |
| 发送方IP地址 |
| 目的方MAC地址 |
| 目的方IP地址 |

图 5-3 ARP 报文格式（语法）

MAC 地址长度：1 个字节，用于规定 MAC 地址字段的长度。例如，以太网 MAC 地址字段一般占 6 个字节（48 位地址）。

IP 地址长度：1 个字节，用于规定 IP 地址字段的长度。例如，IPv4 协议下 IP 地址字段占 4 个字节。

操作：2 个字节，表示报文类型，不同的报文类型代表不同的操作。其中，"1"表示 ARP 请求报文；"2"表示 ARP 响应报文；"3"表示 RARP 请求报文；"4"表示 RARP 响应报文。

发送方 MAC 地址：6 个字节，用于存放发送方 MAC 地址。

发送方 IP 地址：4 个字节，用于存放发送方 IP 地址。

目的方 MAC 地址：6 个字节，用于存放目的方 MAC 地址，对于 ARP 请求报文，该字段为空，一般填写 48 比特位的 0。

目的方 IP 地址：4 个字节，用于存放目的方 IP 地址；对于 RARP 请求报文，该字段为空，一般填写 32 比特位的 0。

发送方在发送 ARP 请求报文时，要填写除目的方 MAC 地址字段之外的其他字段。目的方通过发送 ARP 响应报文来报告自己的 MAC 地址，这时报文中的发送方字段和目的方字段要进行相应的交换。

如果发送方与目的方在同一个网段，则发送方的 ARP 请求报文可直接发送给网络中的任何一个主机，然后通过 ARP 协议直接获得目的方 MAC 地址。如果发送方与目的方在不同网段，则发送给另一网络主机的 IP 分组需要三层网络设备的路由转发。因此，发送方必须先获取本地网段三层网络设备的 MAC 地址，否则无法通过 ARP 协议直接获得目的方 MAC 地址，只能获得本地网段网关的 MAC 地址。

5.2.2 ARP 协议语义

ARP 请求报文：发送方发送 ARP 请求报文，以获取目的方或网关的 MAC 地址。ARP 请求报文在数据链路层被封装为二层广播帧，本地网段的所有主机会在数据链路层接收到该广播帧。

ARP 响应报文：网络中的主机或网关在收到 ARP 请求报文后，只有 ARP 请求报文中封装的目的 MAC 地址与接收方 MAC 地址匹配时,接收方才会发送一个 ARP 响应报文,告诉发送方其想知道的接收方 MAC 地址。ARP 响应报文在数据链路层被封装为二层单播帧。

RARP 请求报文:发送方向本地 RARP 服务器发送 RARP 请求报文,以获取发送方 IP 地址。RAPP 请求报文在数据链路层被封装为二层广播帧,本地网段所有主机会在数据链路层接收到该广播帧。

RARP 响应报文:网络中的主机或网关在数据链路层收到 RARP 请求报文后,只有 RARP 服务器才会发送一个 RARP 响应报文,并按照一定的 IP 地址分配策略,分配一个 IP 地址并通知发送方。RARP 响应报文在数据链路层被封装为二层单播帧。

5.2.3 ARP 协议时序关系

当源主机 A 向目的主机 B 发送一个 IP 分组时,只有知道与对方 IP 地址对应的 MAC 地址,才能在网络上进行帧的传输;源主机 A 构建好一个 IP 分组后,首先将源 IP 地址和目的 IP 地址分别与自己的子网掩码(以下简称掩码)进行与操作,以获得两个网络地址,即源网络地址与目的网络地址;如果这两个网络地址相同,则表明源主机 A 与目的主机 B 在同一个网段。此时,源主机 A 在自己的 ARP 缓存中检查是否存在目的主机 B 的 IP 地址对应的 MAC 地址,如果存在且在生存期内,则不用调用 ARP 协议,直接获取目的主机 B 的 MAC 地址,并将目的主机 B 的 MAC 地址通过网络层与数据链路层接口告知本地数据链路层,用于封装数据帧时填写目的 MAC 地址。如果得到的源网络地址与目的网络地址不同,则表明源主机 A 与目的主机 B 不在同一网段(假设目的主机 B 位于图 5-4 中主机 C 的位置),源主机 A 向目的主机 B 发送的 IP 分组需要通过三层网络设备的路由转发才可以到达。因此,源主机 A 依据目的主机 B 的 IP 地址查询自己的路由表,然后发现要到达目的网络的下一条 IP 地址,该 IP 地址实际上是主机 A 所在网络的网关 IP 地址,此时主机 A 发送一个 ARP 请求报文(二层广播帧),向本网段所有的网络接口询问网关 IP 地址对应的 MAC 地址;本网段所有的网络接口均会接收到该二层广播帧,只要网关满足条件,便发送一个 ARP 应答(二层单播帧),告知主机 A 本地网关的 MAC 地址。

ARP 协议工作网络拓扑结构如图 5-4 所示。

图 5-4 ARP 协议工作网络拓扑结构

在图 5-5 中,如果主机 H1 和主机 H2 不在同一个网段,则 ARP 协议工作方式如下。

(1)如果发送方是主机 H1,要把 IP 分组发送到另一个网络上的主机 H2,这时主机 H1 使用 ARP 协议只能获得本网络上的一个路由器的 MAC 地址,也就是路由器 R1 的 HA3 接口对应的 MAC 地址 MAC3。

（2）如果发送方是路由器 R1，要把 IP 分组转发到另一个网络上的主机 H2，这时路由器 R1 使用 ARP 协议只能获得路由器 R2 的 HA5 接口对应的 MAC 地址 MAC5。

（3）如果发送方是路由器 R2，要把 IP 分组转发到目的方主机 H2，这时路由器 R2 使用 ARP 协议可以直接获得目的方主机 H2 的 HA2 接口对应的 MAC 地址 MAC2。

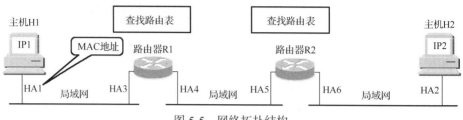

图 5-5　网络拓扑结构

假设在图 5-5 中，主机 H1、路由器 R1 和路由器 R2，以及主机 H2 在不同网络接口下的 IP 地址和 MAC 地址分配如表 5-1 所示。

表 5-1　网络节点地址分配表

网络节点	主机 H1	路由器 R1		路由器 R2		主机 H2
网络接口	HA1	HA3	HA4	HA5	HA6	HA2
IP 地址	IP1	IP3	IP4	IP5	IP6	IP2
MAC 地址	MAC1	MAC3	MAC4	MAC5	MAC5	MAC2

在以上 3 个不同网段，主机 H1 发送给主机 H2 的源 IP 地址、目的 IP 地址，以及不同网段的 ARP 报文在数据链路层的数据帧的目的 MAC 地址和源 MAC 地址变化如表 5-2 所示。

表 5-2　不同网段分组和 ARP 报文在二层数据帧地址变化表

类别	网段 1（HA1 到 HA3）		网段 2（HA4 到 HA5）		网段 3（HA6 到 HA2）	
	源地址	目的地址	源地址	目的地址	源地址	目的地址
IP 分组（IP 地址）	IP1	IP2	IP1	IP2	IP1	IP2
IP 分组（MAC 地址）	HA1	HA3	HA4	HA5	HA6	HA2
ARP 请求报文（MAC 地址）	MAC1	0XFFFFFF…F	MAC4	0XFFFFFF…F	MAC6	0XFFFFFF…F
ARP 应答报文（MAC 地址）	MAC3	MAC1	MAC5	MAC4	MAC2	MAC5

为什么不直接使用 MAC 地址进行通信呢？由于全世界存在着各式各样的网络，它们分别使用不同的 MAC 地址，因此要使这些异构网络能够互相通信，就必须进行非常复杂的 MAC 地址转换工作，但这几乎是不可能的事。连接到因特网的主机都拥有统一的 IP 地址，它们之间的通信就像连接在同一个网络上那样简单方便；调用 ARP 协议来寻找某个路由器或主机的 MAC 地址都是由计算机软件自动进行的，用户是看不见这种调用过程的。

ARP 协议要点：发送方在发送 ARP 请求报文时，除接收方 MAC 地址字段之外的其他字段均要填写。发送方通过发送 ARP 响应报文报告自己的 MAC 地址。如果发送方与接收方在同一网段,则发送方的 ARP 请求报文可直接发送给本网段中的目的主机,并直接获得目的方主机的 IP 地址对应的 MAC 地址;如果发送方与接收方不在同一网段,则发送方实际上是想获取"下一跳"接口的 MAC 地址。从 IP 地址到 MAC 地址的解析是自动进行的,用户并不知道这种地址的解析过程,即该过程是透明的。

RARP 协议功能:无 IP 地址的主机可以通过 RARP 协议利用自己的 MAC 地址来获取相应的 IP 地址;RARP 协议主要通过无盘工作站来获取自己的 IP 地址。RARP 报文格式与 ARP 报文格式相同,其工作原理如下。

(1) 如果一个主机初始化后只有自己的 MAC 地址而没有 IP 地址,则以可通过 RARP 协议发送请求报文来请求 RARP 服务器告知主机的 IP 地址。

(2) 如果发送方以广播方式发送 RARP 请求报文,则发送方 MAC 地址字段和接收方 MAC 地址字段都填入本机 MAC 地址。

(3) RARP 服务器在接收到请求报文后,根据 MAC 地址+IP 地址映射表,给发送方发送一个 RARP 响应报文,然后从接收方 IP 地址字段中带回发送方的 IP 地址。

5.3 ARP 协议发送报文

ARP 协议发送方实现源代码如下。

```
#include <iostream>
#include <pcap.h>
#pragma comment(lib,"wpcap.lib")
#pragma comment(lib,"ws2_32.lib")

using namespace std;
//工业以太网数据帧首部数据结构
typedef struct PhyFrame{
    unsigned char DstMAC[6];            //目的 MAC 地址
        unsigned char SrcMAC[6];        //源 MAC 地址
        unsigned short FrameType;       //协议类型
}frameHeader, * ptrFrameHeader;


//ARP 响应报文数据结构
typedef struct tagArpHeader
{
    unsigned short HardwareType;        //网络类型
        unsigned short ProtocolType;    //上层协议类型
        unsigned char  MACLen;          //MAC 地址长度
        unsigned char  IPLen;           //IP 地址长度
        unsigned short Flag;            //1 表示请求
        unsigned char  SrcMAC[6];       //源 MAC 地址
        unsigned long  SrcIP;           //源 IP 地址
        unsigned char  DstMAC[6];       //目的 MAC 地址
        unsigned long  DstIP;           //目的 IP 地址
```

```cpp
    unsigned char   Padding[18];           //填充数据
}arpHeader, *arpHeader;
void main(int argc,char *argv[])
{
    //填充工业以太网数据帧首部各字段
    int i=0;
    memset(&arpPacket,0,sizeof(arpPacket));
    if(!CheckMAC(argv[4],arpPacket.phyFrame.DstMAC))
        return;
    if(!CheckMAC(argv[2],arpPacket.phyFrame.SrcMAC))
        return;

    //填充ARP协议数据单元首部各字段
    arpHeader = {0};
    arpHeader.FrameType=htons((unsigned short)0x0608);
    arpHeader.HardwareType=(unsigned short)0x0100;
    arpHeader.ProtocolType=(unsigned short)0x0008;
    arpHeader.MACLen=(unsigned char)6;
    arpHeader.IPLen=(unsigned char)4;
    arpHeader. SrcMAC  = "111111111111"
    arpHeader. SrcIP   = inet_addr("192.168.0.1");
    arpHeader. DstMAC  = "222222222222";
    arpHeader. DstIP   = inet_addr("192.168.0.1");
    for(i=0;i<18;i++)
        arpPacket.arpFrame.Padding[i]=0;
    //初始化网络接口相关参数
    pcap_if_t *alldevs;
    pcap_if_t *d,*head=NULL;
    pcap_t *fp;
    char errbuf[PCAP_ERRBUF_SIZE];
    //获取网络接口设备列表
    if(pcap_findalldevs(&alldevs,errbuf)==-1){
        cout<<"Unable to create adapter list!"<<endl;
        return 0;
    }
    //打印网络接口设备列表
    i=0;
    for(d=alldevs;d;d=d->next){
        cout<<++i<<":"<<d->name;
        if(d->description)
            cout<<" "<<d->description<<endl;
    }

    //如果未检测到本地网络接口设备
    if(i==0){
        cout<<"No adapter found!"<<endl;
```

```
        return 0;
   }
   //选择本次通信要使用的网络接口设备
   cout<<"Enter the interface number (1-"<<i<<"):";
   int k;
   cin>>k;
   if(k<1||k>i){
       cout<<"Out of range!"<<endl;
       return;
   }
   for(d=alldevs,i=1;i<k;d=d->next,i++)
       head=d;

   //以混杂模式打开或启动网络接口设备
   if((fp=pcap_open_live(head->name,1000,1,1000,errbuf))==NULL){
       cout<<"Unable to open the adapter!"<<endl;
       pcap_freealldevs(alldevs);
       return;
   }
   //通过网络接口设备发送ARP请求报文
   if(pcap_sendpacket(fp,(unsigned char*)&arpPacket,sizeof((unsigned
           char*)&arpPacket))==-1){
       cout<<"ARP packet send error!"<<endl;
       return 0;
   }
   //ARP响应报文各字段内容
   cout<<"Source IP:"<<argv[1]<<endl;
   cout<<"Source MAC:"<<argv[2]<<endl;
   cout<<"Dest IP:"<<argv[3]<<endl;
   cout<<"Dest MAC:"<<argv[4]<<endl;
   cout<<"ARP packet send success!"<<endl;
}
```

5.4 ARP 协议接收报文

ARP 协议接收方实现源代码如下。

```
#include <stdlib.h>
#include <stdio.h>
#include <pcap.h>
#pragma comment(lib, "wpcap.lib")
#pragma comment(lib, "wsock32.lib")
#pragma comment(lib, "ws2_32.lib")
//定义ARP协议数据单元数据结构
typedef struct arppkt
{
```

```c
    unsigned short hdtyp;          //网络类型
    unsigned short protyp;         //协议类型
    unsigned char hdsize;          //MAC 地址长度
    unsigned char prosize;         //IP 协议地址长度
    unsigned short op;             //操作类型:ARP/RARP
    u_char smac[6];                //源 MAC 地址
    u_char sip[4];                 //源 IP 地址
    u_char dmac[6];                //目的 MAC 地址
    u_char dip[4];                 //目的 IP 地址
}arpp;
int main(int argc,char * argv[] )
{
    struct tm * timeinfo;
    struct tm *ltime;
    time_t rawtime;
    FILE * fp=NULL;
    int result;
    int i=0,inum;
    pcap_if_t * alldevs;                         //指向 pcap_if_t 结构列表的指针
    pcap_if_t * d;
    pcap_t * adhandle;                           //定义捕获数据包句柄
    char errbuf[PCAP_ERRBUF_SIZE];               //错误信息缓冲最小为 256 字节
    u_int netmask;                               //定义掩码
    char packet_filter[]="ether proto \\arp";
    struct bpf_program fcode;
    struct pcap_pkthdr * header;
    const u_char * pkt_data;
    //打开文件
    if((fp=fopen(argv[1],"a"))==NULL)
    {
        printf("打开文件失败!\n");
        exit(0);
    }
    if (argc != 2)             //argc==2,程序名后面有 1 个参数
    {
        printf("程序%s 需要一个文件名参数!\n", argv[0]);
        return -1;
    }
    //获取本地网络接口设备列表
    if (pcap_findalldevs(&alldevs, errbuf) == -1)
    {
        fprintf(stderr,"Error in pcap_findalldevs: %s\n", errbuf);
        exit(1);
    }
    //打印本地网络接口设备列表
    for(d=alldevs; d; d=d->next)
```

```c
{
    printf("%d. %s", ++i, d->name);
    if (d->description)
        printf(" (%s)\n", d->description);
    else
        printf(" 该网络接口描述符不可用\n");
}
if(i==0)
{
    printf("\n找不到网卡！检查是否安装WinPcap.\n");
    return -1;
}
printf("选择对应网卡编号 (1-%d):",i);
scanf("%d", &inum);
if(inum < 1 || inum > i)
{
    printf("\n输入的编号超出范围！\n");
    /*释放网络接口列表*/
    pcap_freealldevs(alldevs);
    return -1;
}
/*选择本次通信需要的网络接口设备*/
i=0;
d=alldevs;
while(i<inum-1)
{
    d=d->next;
    i++;
}
if ( (adhandle= pcap_open_live(d->name,65536, 1,1000,errbuf) ) == NULL)
{
    /*打开网络接口失败*/
    fprintf(stderr,"\n打开失败. %s 不被 winpcap 支持\n",d->name);
    /*释放网络接口设备列表*/
    pcap_freealldevs(alldevs);
    return -1;
}
/*释放网络接口设备列表*/
pcap_freealldevs(alldevs);

//获得掩码
netmask=((sockaddr_in *)(d->addresses->netmask))->sin_addr.S_un.S_addr;
//编译数据包过滤器,仅捕获ARP报文
if(pcap_compile(adhandle,&fcode,packet_filter,1,netmask)<0)
{
    printf("\nUnable to compile the packet filter.Check the syntax.\n");
```

```c
        pcap_freealldevs(alldevs);
        return -1;
}
//设置数据包过滤器过滤规则
if(pcap_setfilter(adhandle,&fcode)<0)
{
    printf("\nError setting the filter.\n");
    pcap_freealldevs(alldevs);
    return -1;
}
//打印每次修改文件的时间
time ( &rawtime );
timeinfo = localtime ( &rawtime );
printf("-------------修改时间: %s",asctime (timeinfo));
fprintf(fp,"-----------修改时间: %s",asctime (timeinfo));
//刷新缓冲流
fflush(fp);
while((result=pcap_next_ex(adhandle,&header,&pkt_data))>=0)
{
    //循环解析ARP报文
    if(result==0)
        continue;
    //解析ARP报文,并将结果输出到屏幕及文件
    arppkt* arph = (arppkt *)(pkt_data +14);
    //输出操作时间
    ltime=localtime(&header->ts.tv_sec);
    printf("时间: %s",asctime (ltime));
    fprintf(fp,"时间: %s",asctime (ltime));
    //输出源IP地址
    printf("源IP: ");
    fprintf(fp,"源IP: ");
    for(i=0;i<3;i++)
    {
        printf("%d.",arph->sip[i]);
        fprintf(fp,"%d.",arph->sip[i]);
    }
    printf("%d\t",arph->sip[3]);
    fprintf(fp,"%d.\t",arph->sip[3]);
    //输出目的IP地址
    printf("目的IP地址: ");
    fprintf(fp,"目的IP地址: ");
    for(i=0;i<3;i++)
    {
        printf("%d.",arph->dip[i]);
        fprintf(fp,"%d.",arph->dip[i]);
    }
```

```c
            printf("%d\t",arph->dip[3]);
            fprintf(fp,"%d\t",arph->dip[3]);
            //输出源MAC地址
            printf("源MAC地址: ");
            fprintf(fp,"源MAC地址: ");
            for(i=0;i<5;i++)
            {
                printf("%x-",arph->smac[i]);
                fprintf(fp,"%x-",arph->smac[i]);
            }
            printf("%x\t",arph->smac[5]);
            fprintf(fp,"%x\t",arph->smac[5]);
            //输出目的MAC地址
            printf("目的MAC地址: ");
            fprintf(fp,"目的MAC地址: ");
            for(i=0;i<5;i++)
            {
                printf("%x-",*(pkt_data+i));
                fprintf(fp,"%x-",*(pkt_data+i));
            }
            printf("%x\t",*(pkt_data+5));
            fprintf(fp,"%x\t",*(pkt_data+5));
            //输出操作类型
            printf("操作类型（ARP/RARP）: ");
            fprintf(fp,"操作类型(ARP/RARP): ");
            if(arph->op==256)
            {
                printf("ARP\t");
                fprintf(fp,"ARP\t");
            }
            else
            {
                printf("RARP\t");
                fprintf(fp,"RARP\t");
            }
            printf("\n");
            fprintf(fp,"\n");
            printf("--------------------------------------\n");
            fprintf(fp,"--------------------------------------\n");
            fflush(fp);
    }
    fclose(fp);
    return 0;
}
```

第 6 章 网络层 IP 协议分析与实践

6.1 引　　言

　　物理层的处理对象为电信号，主要功能及应用包括电信号编码/解码、多路复用，同步方式，以及物理接口特性研究。数据链路层的主要功能是，将不可靠的物理链路转变为可靠的数据链路，其中 LLC 子层主要实现差错控制、流量控制、数据链路管理；MAC 子层主要解决介质的访问控制和成帧问题。网络层应该向上层（传输层）提供"面向连接"还是"无连接"的问题曾引起了长期的争论，争论焦点的实质是，在网络通信中，可靠传输是由网络提供的还是由端系统提供的。起初因特网的基本设计原则是网络越简单越好，最好将网络层的复杂功能都交给端系统处理，而网络层向上层只提供简单灵活的、无连接的、尽最大努力交付的数据报传输和路由服务；端系统主要完成构造、发送、接收和处理 IP 分组，以及网络交换设备对 IP 分组实施存储转发和路由功能；网络层在转发每一个 IP 分组（IP 数据报）时，采用独立路由，所以每个 IP 分组有可能沿着不同路径到达目的终端，因此 IP 分组到达目的节点后会出现乱序。

　　在网络通信中，数据链路层有可能采用不同物理网络（网络接口不同）来传输不同类型的数据帧，但网络层传输协议数据单元（PDU）统一为 IP 分组。网络层不提供 IP 分组的可靠传输（IP 分组在网络传输过程中可能出错、丢失或失序），因此无法保证 IP 分组传输的实时性，如图 6-1 所示。传输层重传的 TCP 数据段在网络层看来都是新的 IP 分组，由于网络层不提供可靠传输，因此不存在重复的 IP 分组传输问题。这使得终端设备（计算机）的设计相对复杂和智能化，网络层传输可靠性主要由端系统上层（传输层）负责，网络尽最大努力交付，越简单越好。这样设计的好处是网络造价降低，运行方式灵活。因特网能够发展到现在的规模，证明了这种设计的正确性。

　　IP 协议是 TCP/IP 协议集的核心协议之一，它提供了"无连接"的 IP 分组传输和互联网的路由服务。IP 协议的基本任务是通过互联网传输 IP 分组（或 IP 数据报），各 IP 分组独立传输。主机上的 IP 分组基于数据链路层向传输层提供传输服务，IP 协议从源主机的传输层获取数据，再通过物理网络传送给目的主机的网络层。IP 协议不保证传输的可靠性，在网络资源不足的情况下，网络设备可能会丢弃某些 IP 分组，同时 IP 协议也不检查被数据链路层丢弃的数据帧。

　　在传送数据时，上层协议将数据传送给 IP 协议，IP 协议将数据封装为 IP 分组后通过网络接口发送出去。如果目的主机直接连在本地网络中，则网络可直接将 IP 分组传送给本地网络中的目的主机；如果目的主机是在远程网络上的，则 IP 协议需要将 IP 分组传送给本地路由器，然后由本地路由器将 IP 分组传送给下一个路由器。这样，一个 IP 分组通过网络从一个 IP 实体传送到另一个 IP 实体，直至到达目的地为止。

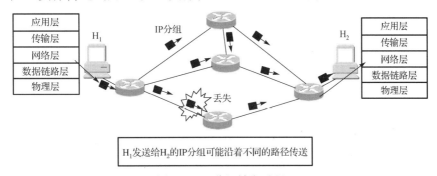

图 6-1　IP 分组转发过程

IP 协议主要包括以下功能。

（1）IP 分组的生成、发送、接收和处理。

（2）数据段的分片与组装。

（3）分组转发：每个 IP 分组（包括分片）依据目的 IP 地址查找路由表中的路由信息，再进行独立的路由选择。

TCP/IP 协议集体系结构如图 6-2 所示。

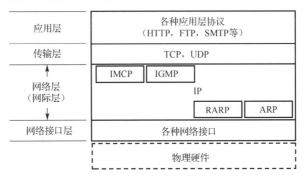

图 6-2　TCP/IP 协议集体系结构

网络层使用的 4 个协议如下。

（1）网际控制报文协议（Internet Control Message Protocol，ICMP）。

（2）因特网组管理协议（Internet Group Management Protocol，IGMP）。

（3）地址解析协议（Address Resolution Protocol，ARP）。

（4）逆地址解析协议（Reverse Address Resolution Protocol，RARP）。

在因特网中，每个网络设备（或网络接口）至少有一个 IP 地址，端系统至少有一个网络接口，每个网络接口分配一个 IP 地址；主机和路由器可以有多个网络接口，每个网络接口有一个 IP 地址。三层交换机比较特殊，它的每个网络接口构成一个网段，一个网段虚拟出一个网关，该网关占用一个 IP 地址。

IP 地址与 MAC 地址的关系：MAC 地址写在网络适配器内存中，一般不能改变；IP 地址可灵活配置，方便用户改变和管理，适用范围为整个互联网；MAC 地址的作用范围为虚拟局域网内部，因此其只需要在虚拟局域网内部唯一即可。

6.2 IP 协议工作原理

6.2.1 IP 协议语法

一个 IP 分组由首部和数据两部分组成，首部又由固定部分和可变部分组成，其中，固定部分的长度是固定的，为 20 字节，所有 IP 分组必须具有固定部分（分片也是一个独立的 IP 分组）；可变部分的长度可变（0～40 字节），由选项字段和填充字段两部分组成，如图 6-3 所示。计算机网络在 IP 层只能处理 IP 分组；路由器接收到一个 IP 分组，根据 IP 分组的目的 IP 地址网络号进行路由选择。在物理网络数据链路层，用户看不到 IP 分组，只能看见数据帧，因为 IP 层抽象的互联网屏蔽了下层复杂的细节；在抽象的网络层上讨论问题，能够使用统一的、抽象的 IP 地址；而在实际物理网络中，当主机与主机之间或主机与路由器之间通信时，数据帧的 MAC 地址（源、目的）在不同物理网络上均发生了变化，因为源主机在发送数据帧时，需要封装 MAC 地址（源、目的），路由器在转发 IP 分组时，在网络接口数据链路层需要改变数据帧的 MAC 地址（源、目的）。

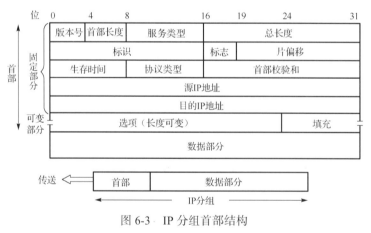

图 6-3 IP 分组首部结构

（1）版本号：占 4 比特位，表示 IP 协议的版本号，目前的 IP 协议版本号为 4（IPv4：0100）。

（2）首部长度：占 4 比特位，可表示的最大数值是 15 个单位（一个单位为 4 字节），所以 IP 分组的首部长度的最大值是 60 字节，其中固定长度 20 字节；可变部分长度：0～40 字节；因此，IP 分组首部长度范围为 20～60 字节。

（3）服务类型（Type Of Service，TOS）：占一个字节，具体内容如图 6-4 所示。

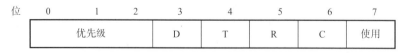

图 6-4 IP 分组首部 TOS 字段结构

优先级：占 3 比特位，共 8 个优先级别，其中 0 表示最小优先级，7 表示最大优先级。

在图 6-4 中，D 表示低延迟（Delay）；T 表示高吞吐量（Throughput）；R 表示高可靠性（Reliability），被路由丢弃概率较小；C 表示选择代价更小路由。

TOS 只是源节点用户要求和填写的，对网络并不是强制的，在路由器进行路由选择等处理时仅作为参考。

（4）总长度：占16比特位，包括IP分组首部+数据的长度，单位为字节，因此IP分组的最大长度为65535字节，一般大小变化为578～1500字节；实际上，IP分组总长度必须不超过数据链路层最大传输数据单元（MTU）（一个完整数据帧的数据部分长度）。目前，该总长度大小已经足够，在将来的高速网络中，数据帧MTU有可能大于65535字节；由于分片的原因，每个分片也是一个完整的IP分组，每个分片总长度指该IP分片首部+数据两部分长度。

（5）标识（Identification）：占16比特位。

源节点每产生一个新的 IP 分组，通过计数器加 1，作为该 IP 分组的标识，而不是序号，因为 IP 协议提供"无连接"通信，所以不存在按序接收的问题；注意，一个分组不同分片具有相同标识，其中与分片组装有关的字段主要包括：标识+标志+片偏移。

（6）标志（Flag）：占 3 比特位，目前只使用两个比特（0+DF+MF）。

DF（Don't Fragment）：DF=1 表示不允许分片，DF=0 表示允许分片。

MF（More Fragment）：MF=1 表示后面还有分片；MF=0 表示这是分片后最后一个分组。

（7）片偏移：占 12 比特位，针对大的 IP 分组，由于需要进行分片，因此该字段主要指出 IP 分组的数据部分在分片后，某一个分片（或分组）的数据部分的第一个字节在原始数据中的相对位置；片偏移字段的值以 8 个字节为偏移单位。

（8）生存时间（Time To Live，TTL）：占 8 比特位，一般由源节点的操作系统设置该字段的值，不同操作系统设置的初始值不同，其目的是丢弃网络中存在很长时间的分组。源节点一般将 TTL 字段设置为 64，最大 255，每经过一个路由器，在转发前 TTL 减 1。当 TTL=0 时，路由器丢弃该分组，同时向源节点发送一个 ICMP 超时信息。当网络中存在环路，或者 IP 分组找不到确切路由而一直使用默认路由等情况时，会发生 TTL 超时，此时 IP 协议调用 ICMP 协议，并向源节点发送 TTL 超时的 ICMP 差错报告报文。

（9）协议类型：占 8 比特位，可指出 IP 分组携带的数据来自何种上层协议，以便目的主机在 IP 层将数据部分交付给相应上层协议进程进行处理。协议号由国际组织 IANA 负责分配。

（10）首部校验和：占 16 比特位，只校验 IP 分组首部，不校验数据部分；为了减少 IP 层处理 IP 分组的时间，数据差错控制由端系统传输层完成，IP 协议不采用 CRC 校验码，而采用简单校验和计算方法；简单校验和计算方法是将 IP 分组首部按照 16 比特位对齐（按照实际情况补 0），校验和字段先填写 16 比特位的 0，然后按照二进制加法计算，以计算的结果（最高位进位丢弃）取反作为简单校验和计算的结果填写在该字段。

接收方校验的正确性采用和发送方类似的计算方法，即先将 IP 分组首部按照 16 比特位对齐（按照实际情况补 0），再按照二进制加法计算，以计算的结果（最高位进位丢弃）取反作为检验结果。如果计算结果为 16 比特位的 0，则说明 IP 分组的首部在网络传输时没有发生错误，否则认为 IP 分组的首部发生了错误，此时，IP 协议会丢弃该 IP 分组，并调用 ICMP 协议向源节点发送 IP 分组的首部出错的 ICMP 差错报告报文。在 IP 分组从一个路由器转发给另一个路由器前，由于 TTL 发生变化，片偏移也有可能发生变化，因此 IP 协议需要重新计算 IP 分组的首部校验和。

（11）源 IP 地址和目的 IP 地址：各占 4 字节；在 IP 分组转发路径上，这两个字段的值一直保持不变（除了 NAT 转化）。

（12）选项：占 0～40 字节，用于网络控制、测量和安全；可提供各种选项服务，如时间戳选项、源路由选项等；IP 分组首部可变部分在增加 IP 协议功能的同时，使得 IP 分组首部长度可变，增加了每一个路由器处理分组的开销，但实际上这些选项现在很少使用。

（13）填充：长度可变，保证 IP 分组首部以 32 比特位边界对齐。

6.2.2 IP 协议语义

IP 广播分组：IP 分组的目的 IP 地址可以是广播地址。广播地址分为两类，一类是直接广播地址，其特点是包含一个有效的网络号和一个全"1"的主机号，如 202.163.30.255/24，表示一个 C 类直接广播地址；另一类是受限广播地址，受限的广播地址为 255.255.255.255。如果一个 IP 分组的目的 IP 地址是一个直接广播地址，则 IP 分组只能在局域网内传播，无法跨越三层网络设备；如果一个 IP 分组的目的 IP 地址是一个受限广播地址，则 IP 分组传输到三层网络设备后，三层网络设备依据受限广播地址查找路由表进行转发，直至到达目的网络为止。

IP 单播分组：如果一个 IP 分组的目的 IP 地址是一个正常的单播地址，则 IP 分组传输到三层网络设备后，三层网络设备依据目的 IP 地址（依据目的网络地址）查找路由表进行转发，直至到达目的网络为止。

IP 组播分组：如果一个 IP 分组的目的 IP 地址是一个组播地址，则该 IP 分组依据组播路由信息进行转发，最终到达目的组播组所在的所有成员的节点。

IP 分片：当传输层交付给网络层 IP 协议的协议数据单元比较大时，IP 协议需要将传输层协议数据单元分片，而且每一个分片有一个完整的 IP 首部，构成一个 IP 分组，每个 IP 分组在网络中拥有独立路由，可能沿着不同路径到达目的节点，并在目的节点网络层重新组装，将数据部分按照协议类型交付给上层协议进行下一步处理。

6.2.3 IP 协议时序关系

1）IP 分组分片机制

在各种物理网络中，如以太网、令牌环网等网络都有最大帧长限制，因此为了使较大的 IP 分组能以适当的大小在物理网络上传输，IP 协议要先根据物理网络允许的最大帧长对传输层提交的协议数据单元进行长度检查，必要时要把传输层提交的协议数据单元分成若干个片，并以 IP 分片的形式发送到网络中。

在传输层协议数据单元进行分片时，每个分片都要加上 IP 首部，形成一个独立的 IP 分组。与分片相关的字段如下。

（1）标识（ID）：IP 分组的唯一标识。分片后传送的 IP 分组均有相同的标识。

（2）报文长度：对每一个已分片的 IP 分组都要重新计算其报文长度。

（3）片偏移：每一个已分片的 IP 分组都要表明它在原始数据中的位置，并用 8 字节的倍数来表示。

（4）标志（Flag）：如果是未分片的 IP 分组，该标志位置 0；如果是已分片的 IP 分组，除了最后一个分片将该标志位置 0，其他的分片都将该标志位置 1。

为了使较大的 IP 分组以适当的大小在物理网络上传输，IP 协议首先要根据物理网络允许的最大帧长对上层协议提交的协议数据单元长度进行检查，必要时要将其分成若干个分组（分片）发送；分片的对象为上层协议数据单元（分组的数据部分），不包括 IP 首部。由于每个分片独立发送，独立路由，可能造成在接收方出现乱序的情况，因此在接收方需要时应重新排序组装。

传输层交付给网络层协议数据单元的长度为 3800 个字节，IP 协议依据数据链路层 MTU 大小，将其分为 3 个分片，这 3 个分片的数据部分的长度分别是 1000 字节、1000 字节和 800 字节，片偏移字段的值分别为 0、175 和 350，如图 6-5 所示。如果此时系统分配的分组标识为 12345，则这 3 个分片的 IP 首部与分片相关字段的值如表 6-1 所示。

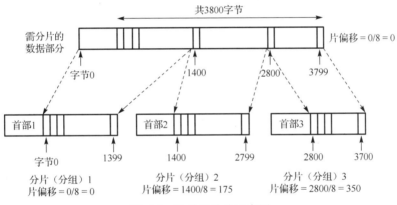

图 6-5　IP 分组分片示意图

表 6-1　IP 分片相关信息表

	总长度	标识	标志 MF	标志 DF	片偏移
原始段	3820 字节	12345	0	0	0
分组 1	1420 字节	12345	1	0	0
分组 2	1420 字节	12345	1	0	175
分组 3	1020 字节	12345	0	0	350

从表 6-1 可以看出，一个大的上层协议数据单元被分为 3 个分片在网络上传输，这 3 个分片作为独立的 IP 分组，在 IP 网络上独立路由，可能沿着不同路径到达目的方。如果表 6-1 中的分组 2 在传输过程中由于某一链路 MTU 要求需要进一步分片，则将分组 2 的数据部分再分为 2 个分片，每个分片的数据部分的长度分别为 800 字节和 600 字节，而这两个分片对应的片偏移分别为 175 和 275，如图 6-6 所示。此时，从传输层交付给网络层

的长度为 3800 字节的协议数据单元被分为 4 个分片，这 4 个分片是 4 个独立的 IP 分组，与这 4 个分组首部分片相关的字段值如表 6-2 所示。

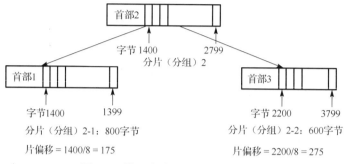

图 6-6　第 2 个分片继续分片示意图

表 6-2　第二个分组继续分片后与分片相关字段值

	总长度	标识	标志		片偏移
			MF	DF	
原始段	3820 字节	12345	0	0	0
分组 1	1420 字节	12345	1	0	0
分组 2-1	820 字节	12345	1	0	175
分组 2-2	620 字节	12345	1	0	275
分组 3	1020 字节	12345	0	0	350

假设一个应用程序使用 UDP 协议通信，到了 IP 层将 UDP 用户数据报分为 4 个分片进行传输，结果前两个分片丢失，后两个分片到达目的节点；一段时间后，应用程序启动可靠通信机制重发 UDP 用户数据报，到了 IP 层将 UDP 用户数据报继续分为 4 个分片进行传输，结果前两个分片到达目的节点，后两个分片丢失。假设目的节点第一次收到的后两个分片仍然保存在目的节点缓存中，那么目的节点是否可将两次传输的 8 个分片中，第一次传输的后两个分片和第二次传输的前两个分片组装成一个大的 UDP 用户数据报呢？答案是无法重新组装，因为两组 IP 分组（分片）的标识字段值不同。

2）IP 分组重新组装机制

在网络中，分片的各 IP 分组会进行独立的传输，它们在经过中间路由器转发时可能选择不同的路由。这样，到达目的主机的 IP 分片的顺序与发送的顺序可能不一致。因此，目的主机上的 IP 协议必须根据 IP 分组中的相关字段（标识、报文长度、片偏移及标志等）将分片的各数据字段重新组装成完整的原始数据，然后再交付给上层协议。

在进行分片重新组装时，各 IP 分组除应具有相同的标识之外，还应具有相同的上层协议号、源 IP 地址和目的 IP 地址，并且在一定的时间内要全部到齐。IP 协议将满足上述条件的 IP 分组按片偏移值大小排队，然后将数据部分组装成一个完整的数据单元，最后将组装好的数据单元按上层协议号提交给上层协议。分组（分片）的组装条件如下。

（1）在进行数据分组重新组装时，各 IP 分组应具有相同标识，并按片偏移值大小排队。

（2）各 IP 分组应具有相同的上层协议号、源 IP 地址和目的 IP 地址，并在一定的时间内全部到齐。

（3）判断分片丢失依据：一是 MF=1 的分片必须到达，表明最后一个分片已经接收到；通过分片的片偏移值*8 +IP 首部（总长度-首部长度）是否等于下一个分片的片偏移值*8 来判断是否有中间分片丢失。

如果个别分组（分片）首部有错误，或者有个别分组没有在规定时间到达，那么会造成数据组装错误，此时 IP 协议调用 ICMP 协议向源节点发送 ICMP 分片组装超时差错报告报文。

IP 分组重新组装的具体方法如下。

（1）各 IP 分片按照一定时间到达接收方，利用首部长度、总长度和片偏移字段可推测出是否有 IP 分片丢失；计算方法：分片的片偏移值*8 + 数据长度（总长度-首部长度）= 下一个分片的片偏移值*8。

（2）按片偏移值大小顺序，将多个分片组装成一个完整的数据。

（3）最后将组装好的数据单元按上层协议号提交给上层协议。

如果接收方传输层检测到交付的数据有错误，则发送方传输层需要启动可靠通信机制重发该 TCP 数据段；如果一个网络层分片丢失，则会造成分片组装失败，从而引起发送方 TCP 协议的重发定时器超时，因此该分片对应的 TCP 数据段需要重发一次。

3）网络层路由选择机制

路由选择是 IP 协议非常重要的功能之一。在 IP 协议中，可以采用两种路由选择机制，一种是源路由选择机制，由发送端在选项字段指定发送的 IP 分组源路由信息，三层交换设备接收到该分组后，依据源路由信息进行转发即可。另一种是基于路由表进行路由转发，一个三层交换设备接收到一个 IP 分组后，选择得到该 IP 分组的目的 IP 地址，利用目的 IP 地址与路由表中第二项（掩码或前缀长度）计算得到目的网络地址，将计算得到的目的网络地址与路由表提供的网络地址进行匹配，按照主机路由、优先级、代价最小和最长匹配的选择策略，选择一条最佳路由信息进行转发。

4）源节点 IP 分组发送机制

当源节点 IP 协议收到上层协议要求发送协议单元时，如果上层协议已指定了源路由信息，则按指定的路由转发 IP 分组；如果上层协议未指定源路由信息，则 IP 协议以 IP 分组中目的 IP 地址为关键字来搜索路由表中的路由。如果未找到任何路由，则说明目的不可达，应向上层协议报告错误信息。对于已确定的路由信息，无论是由上层协议指定源路由，还是从路由表中找到路由信息，如果该路由是直接可达的（源主机和目的地址在同一网络中），则 IP 协议调用 ARP 协议获得目的 MAC 地址，并将 IP 分组提交给对应的接口数据链路层进行下一步成帧处理；如果该路由不是直接可达的，而是需要路由转发的，则 IP 协议需要依据路由信息得到下一跳 IP 地址，并调用 ARP 协议获得下一跳 IP 地址对应的 MAC 地址，最后将 IP 分组提交给对应接口数据链路层进行下一步成帧处理。

5）中间节点 IP 分组转发机制

当中间节点（如路由器或三层交换设备）IP 协议进程接到一个 IP 分组后，首先计算 TTL-1 的值并判断其是否为 0，如果为 0 则丢弃，并调用 ICMP 协议向源节点发送 TTL 超时差错报告报文，然后计算校验和，如果校验和出错，则调用 ICMP 协议向源节点发送 IP 首部出错报告报文；如果校验和没有问题，则缓存在接收队列。然后另一个后台进程从接收队列按照 FIFO 策略读取一个 IP 分组，如果该分组提供了源路由信息，则按照源路由信息转发，在转发之前应调用 ARP 协议获取下一跳 IP 地址对应的 MAC 地址；如果该 IP 分组未提供源路由信息，则依据目的 IP 地址查路由表来获得下一跳 IP 地址。如果下一跳 IP 地址还没有到达目的网络，则调用 ARP 协议获取下一跳 IP 地址对应的 MAC 地址，否则可以通过 ARP 协议直接获得目的节点的 MAC 地址。在得到 MAC 地址后，该分组转发给输出接口数据链路层进行下一步的成帧处理。路由表结构如表 6-3 所示。

表 6-3　路由表结构

IP 地址	掩码	下一跳	代价（Metric）	优先（Preference）
11.0.0.0	255.0.0.0	8.8.8.9	30	100（RIP）
11.168.0.0	255.255.0.0	12.8.8.9	10	40（OSPF）
11.168.1.0	255.255.255.0	13.8.8.9	40	20（静态）
11.168.2.0	255.255.255.0	Direct	20	0
0.0.0.0	0.0.0.0	14.8.8.9	200	默认

6）目的节点接收 IP 分组机制

当目的节点收到 IP 分组后，首先比较 IP 分组中的目的 IP 地址与本机 IP 地址是否匹配，如果匹配，则检查校验和字段是否正确，如果正确，则去掉 IP 首部，将 IP 分组的数据部分依据协议类型交付给上层协议进行下一步处理。如果目的节点收到的是分片，则需要依据组装机制，将各分片中数据部分的内容组装成一个完整的上层协议数据单元，再依据协议类型将该完整的协议单元交付给上层协议进行下一步处理。

7）IP 分组选项功能机制

IP 分组的选项字段中提供了若干选项，主要用于网络控制和测试。选项功能作为 IP 协议的组成部分，在所有的 IP 协议实现中都是不可缺少的。目前 IPv4 协议提供的主要选项功能有源路由选项、记录路由选项和时间戳选项。

（1）源路由指转发 IP 分组的路由信息是由源主机指定的，而不是由 IP 协议通过路由表确定的。通过源路由选项功能，可以使一个 IP 分组沿着提前规定好的路径传输，可以用于测试某特定网络的吞吐率，也可以使 IP 分组绕开出错的网络等。源路由选项有两类：一是限制源路由，二是自由源路由。限制源路由给出的是一条完整的路径，并规定分组经过的路径上的每一个路由器的地址信息，而且在网络传输过程中，IP 分组经过路由器的顺序不可改变。自由源路由给出的不是完整路径，而是精简路径，必要时要由 IP 协议为其补充路径。

（2）记录路由选项是用来记录 IP 分组从源主机到目的主机经过的路径上各路由器的 IP 地址的。记录地址的区域大小是由源主机预先分配并初始化的（最多可记录 10 个 IP 地址），如果预先分配的区域不足以记录下全部地址，则 IP 协议将放弃记录余下地址。此外，只有在源主机、目的主机都同意，并且需要在路径上路由器支持的情况下，记录路由选项才是有效的。利用该选项可得到源路由信息。

（3）时间戳是指当 IP 分组经过一个路由器时，路由器记录的接收时间。通过时间戳选项可以获得 IP 分组在互联网中的时域参数，可用于分析网络的吞吐量、拥塞情况、负载情况等。时间戳选项中的时间以千分之一秒为单位。由于各路由器中的时钟并不严格同步，因此时间戳只是一种大致的参考值。此外，在选择时间戳选项时，可以设置成时间和地址同时记录，默认设置为只记录时间。

6.3 IP 协议发送 IP 分组

IP 分组发送方实现源代码如下。

```
#include "Winsock2.h"
#include "stdio.h"
#include "Ws2tcpip.h"
#pragma comment(lib,"ws2_32.lib")
//定义 IP 分组首部数据结构
typedef struct   tagIPHeader
{
    unsigned char ip_verlen;            //4 位首部长度+4 位版本号
    unsigned char ip_tos;               //8 位服务类型
    unsigned short ip_totallength;      //16 位总长度（字节）
    unsigned short ip_id;               //16 位标识
    unsigned short ip_offset;           //3 位标志+片偏移
    unsigned short ip_ttl;              //8 位 TTL
    unsigned short ip_protocol;         //8 位协议类型(TCP, UDP 或其他)
    unsigned short ip_checksum;         //16 位 IP 分组首部简单校验和
    unsigned long ip_srcaddr;           //32 位源 IP 地址
    unsigned long ip_destaddr;          //32 位目的 IP 地址
}IPHeader, *ptrIPHeader;

int main()
{
    //定义和初始化变量
    const int BUFFER_SIZE = 1000;
    char ip_buffer[BUFFER_SIZE];
    Unsigned short checkBuffer[65535];
    const char *ipData = "create IP packet and send!";

    //初始化 Windows Socket DLL
    WSADATA wsd;
    if (WSAStartup(MAKEWORD(2, 2), &wsd) != 0)
    {
        printf("WSAStartup() failed: %d ", GetLastError());
        return -1;
    }
    //创建 Raw Socket
    SOCKET rawSocket = WSASocket(AF_INET, SOCK_RAW, IPPROTO_IP, NULL, 0,
```

```
            WSA_FLAG_OVERLAPPED);
if (rawSocket == INVALID_SOCKET)
{
    printf("WSASocket() failed: %d ", WSAGetLastError());
    return -1;
}

//设置发送超时时间
Int timeout = 1500;
Setsockopt(RawSocket, SOL_SOCKET, SO_SNDTIMEO, (char *)&timeout,
        sizeof(timeout));

//设置首部控制选项，使用IP_HDRINCL
DWORD bOption = TRUE;
int retResult= setsockopt(s, IPPROTO_IP, IP_HDRINCL, (char*) &bOption,
        sizeof(bOption));
if (retResult == SOCKET_ERROR)
{
    printf("setsockopt(IP_HDRINCL) failed: %d ", WSAGetLastError());
    closesocket(rawSocket);
    WSACleanup();
    return -1;
}
//填充IP分组首部各字段
IPHeader = {0};
IPHeader.ip_version = (0x04<<4|sizeof(IPHeader)/4);
IPHeader.ip_tos = 0;
IPHeader.ip_totallength = sizeif(IPHeader) + sizeof (ICMPHeader) +
        sizeof(icmp_data);
IPHeader.ip_id = 0;
IPHeader.ip_ttl = 64;
IPHeader.ip_protocol=IPPROTO_ICMP;
IPHeader.ip_checksum = 0
IPHeader.ip_srcaddr = inet_addr("192.168.0.1");
IPHeader.ip_dstaddre = inet_addr("192.168.0.2");

//计算IP分组首部简单校验和字段
Memset(checkBuffer, 0, 65535);
Memcpy(checkBuffer, &IPHeader,sizeof(IPHeader));
Memcpu(checkBuffer + sizeif(IPHeader), & ipData,sizeof(ipDara));
IPHeader .checksum = checksum(checkBuffer, sizeif(IPHeader) +
+ sizeof(ipData));
//构建待发送的IP分组
Memset(ip_buffer, 0, 1000);
Memcpy(ip_buffer, &IPHeader, sizeof(IPHeader));
Memcpy(ip_buffer + sizeof(IPHeader),&ipData,sizeof(ipData));
```

```cpp
    //填充Socketaddr_in数据结构
    sockaddr_in  remote_addr;
    remote_addr.sin_family = AF_INET;
    remote_addr.sin_port = htons(8000);
    remote_addr.sin_addr.s_addr = inet_addr("192.168.0.2");
    //发送IP分组
    int iSendSize = size0f(IPHeader) + sizeof(ipData);
    int  retResult = sendto(rawSocket, buffer, iSendSize, 0,
    (struct sockaddr *) &remote_addr, sizeof(remote_addr));
    if (retResult == SOCKET_ERROR)
        printf("sendto() failed: %d ", WSAGetLastError());
    else
        printf("sent %d bytes ", retResult);
    //关闭Socket，释放资源
    closesocket(rawSocket);
    WSACleanup();
    return   true;
}
```

6.4 IP 协议接收 IP 分组

IP 分组接收方实现源代码如下。

```cpp
#include "stdafx.h"
#include "winsock2.h"
#include "ws2tcpip.h"
#include "iostream.h"
#include "stdio.h"
#include <string>
using namespace std;
#pragma comment(lib,"ws2_32.lib")
//构建IP分组首部数据结构
typedef struct _IP_HEADER
{
 union
 {
  BYTE Version;              //版本号
  BYTE HdrLen;               //首部长度
 };
 BYTE ServiceType;            //服务类型
 WORD TotalLen;               //总长度
 WORD ID;                     //标识
 union
 {
  WORD Flags;                //标志
  WORD FragOff;              //片偏移
```

```c
    };
    BYTE TimeToLive;           //TTL
    BYTE Protocol;             //协议类型
    WORD HdrChksum;            //首部校验和
    DWORD SrcAddr;             //源IP地址
    DWORD DstAddr;             //目的IP地址
    BYTE Options;              //选项
}IP_HEADER;

int main(int argc,char *argv[])
{
 if(argc!=1)
 {
  printf("usage error!\n");
  return -1;
 }

 WSADATA wsData;
 //初始化失败，程序退出
 if(WSAStartup(MAKEWORD( 2, 2 ),&wsData)!=0)
 {
  printf("WSAStartup failed\n");
  return -1;
 }
//建立Raw Socket
 SOCKET sock;
 if((sock=socket(AF_INET,SOCK_RAW,IPPROTO_IP))==INVALID_SOCKET )
 {
  printf("create socket failed!\n");
  return -1;
 }
 BOOL flag=TRUE;
 //设置IP分组首部操作选项，其中标志字段设置为true，用户可以亲自对IP分组首部进行处理
 if(setsockopt(sock,IPPROTO_IP,IP_HDRINCL,(char *)&flag,sizeof(flag))==
      SOCKET_ERROR)
 {
  printf("setsockopt failed!\n");
  return -1;
 }
 char hostName[128];
 if(gethostname(hostName,100)==SOCKET_ERROR)
 {
  printf("gethostname failed\n");
  return -1;
 }
```

```cpp
//获取本地IP地址
hostent *pHostIP;
if((pHostIP=gethostbyname(hostName))==NULL)
{
 printf("gethostbyname failed\n");
 return -1;
}
//填写SOCKADDR_IN结构
sockaddr_in addr_in;
addr_in.sin_addr=*(in_addr *)pHostIP->h_addr_list[0];
addr_in.sin_family=AF_INET;
addr_in.sin_port=htons(6000);
//将Raw Socket绑定到本地网卡上
if(bind(sock,(PSOCKADDR)&addr_in,sizeof(addr_in))==SOCKET_ERROR)
{
 printf("bind failed");
 return -1;
}
DWORD dwValue=1;
//设置SOCK_RAW为SIO_RCVALL，以便接收所有的IP分组
#define IO_RCVALL _WSAIOW(IOC_VENDOR,1)
DWORD dwBufferLen[10];
DWORD dwBufferInLen=1;
DWORD dwBytesReturned=0;
if(WSAIoctl(sock,IO_RCVALL,&dwBufferInLen,sizeof(dwBufferInLen),
   &dwBufferLen,sizeof(dwBufferLen),&dwBytesReturned,NULL,NULL)==SOCKET_ERROR)
{
 printf("ioctlsocket failed\n");
 cout<<GetLastError()<<endl;
 return -1;
}
//设置接收数据包的缓存区长度
#define BUFFER_SIZE 65535
 char buffer[BUFFER_SIZE];
 printf("开始解析经过本机的IP数据包\n");
 while(true)
 {
  int size=recv(sock,buffer,BUFFER_SIZE,0);
  if(size>0)
  {

      IP_HEADER ip=*(IP_HEADER *)buffer;

      //解析版本信息
      BYTE version;
      getVersion(ip.Version,version);
```

```cpp
version=ip.Version>>4;
printf("版本=%d\r\n",version);

//解析IP分组首部长度
BYTE headerLen;
getIHL(ip.HdrLen,headerLen);
headerLen=(ip.HdrLen & 0x0f)*4;
printf("头长度=%d(BYTE)\r\n",headerLen);

//解析服务类型
fprintf(file,"服务类型=%s,%s\r\n",parseServiceType_getProcedence(ip.ServiceType),
parseServiceType_getTOS(ip.ServiceType));
BYTE b=(ip.ServiceType>>1)&0x0f;
string stem;
string stem1;
switch(b>>5)
{
case 7:
    stem1 = "Network Control";
    break;
case 6:
    stem1 = "Internet work Control";
    break;
case 5:
    stem1 = "CRITIC/ECP";
    break;
case 4:
    stem1 = "Flash Override";
    break;
case 3:
    stem1 = "Flash";
    break;
case 2:
    stem1 = "Immediate";
    break;
case 1:
    stem1 = "Priority";
    break;
case 0:
    stem1 = "Routine";
    break;
default:
    stem1 = "Unknown";
}
b=(ip.ServiceType>>1)&0x0f;
switch(b)
{
case 0:
    stem1 = "Normal service";
    break;
```

```cpp
        case 1:
            stem1 = "Minimize monetary cost";
            break;
        case 2:
            stem1 = "Maximize reliability";
            break;
        case 4:
            stem1 = "Maximize throughput";
            break;
        case 8:
            stem1 = "Minimize delay";
            break;
        case 15:
            stem1 = "Maximize security";
            break;
        default:
            stem1 = "Unknown";
    }
    printf("服务类型=%s,%s\r\n",stem1.c_str(),stem.c_str());
    //解析数据包长度
    printf("数据报长度=%d(BYTE)\r\n",ip.TotalLen);
    //解析数据包 ID
    printf("数据报 ID=%d\r\n",ip.ID);
    //解析标志
    BYTE DF,MF;
    //getFlags(ip.Flags,DF,MF);
    DF=(ip.Flags>>14)&0x01;
    MF=(ip.Flags>>13)&0x01;
    printf("分段标志 DF=%d,MF=%d\r\n",DF,MF);
    //解析片偏移
    WORD fragOff;
    //getFragOff(ip.FragOff,fragOff);
    fragOff=ip.FragOff&0x1fff;
    printf("片偏移值=%d\rn",fragOff);
    //解析 TTL
    printf("生存期=%d\r\n",ip.TimeToLive);
    //解析协议
    printf("协议=%s\r\n",version=b>>4);
    //解析首部校验和
    printf("头校验和=0x%0x\r\n",ip.HdrChksum);
    //解析源 IP 地址
    printf("源 IP 地址=%s\r\n",inet_ntoa(*(in_addr *)&ip.DstAddr));
    //解析目的 IP 地址
    printf("目的 IP 地址=%s\r\n",inet_ntoa(*(in_addr *)&ip.DstAddr));
 printf("-------------------------------------------------\r\n");
    }
   }
       return 0;
}
```

第 7 章　网络层 ICMP 协议分析与实践

7.1　引　　言

　　IP 协议提供了无连接的 IP 分组或数据报传送服务。如果 IP 分组在传送过程中发生了差错或意外情况，如分组目的不可达、分组在网络中的滞留时间超过其生存期、中转节点或目的节点主机因缓存不足而无法接收或处理等，则网络层需要通过一种通信机制，向源节点报告差错情况，以便源节点对差错进行相应的处理。ICMP 协议正是提供这类差错报告服务的协议。ICMP 协议在 IP 层加入了一类特殊用途的报文机制，以辅助 IP 协议报告差错的需求。

　　ICMP 协议是网络层协议的一部分，其产生的报文要通过 IP 协议封装为 IP 分组，然后发送出去。ICMP 协议有多种报文类型，因此可以提供多种服务，如测试目的可达性和状态、报告目的不可达类型、数据报流量控制、路由重定向、检查循环路由或超长路由、报告差错分组首部信息等。

7.2　ICMP 协议工作原理

7.2.1　ICMP 协议语法

　　每个 ICMP 报文都是作为 IP 分组的数据部分在网络间传送的。ICMP 报文格式如图 7-1 所示。

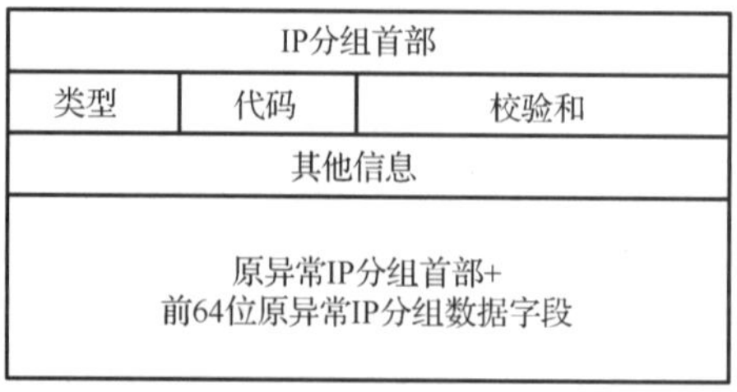

IP分组首部		
类型	代码	校验和
其他信息		
原异常IP分组首部+ 前64位原异常IP分组数据字段		

图 7-1　ICMP 报文格式

　　在图 7-1 中，类型字段占 1 个字节，指出 ICMP 报文的类型；代码字段也占 1 个字节，提供关于报文类型的相关信息，即每种 ICMP 报文类型下的子类型；校验和字段占 2 个字节，校验范围为 ICMP 报文首部+数据字段，数据字段主要包含错误分组 IP 首部+该分组前 64 比特位数据（主要包括传输层首部重要信息，特别是端口号，发送序号）。ICMP 报文类型及含义如表 7-1 所示。

表 7-1 ICMP 报文类型及含义

报文	类型	含义
差错报告报文	3	目的不可达
	11	超时
	12	参数出错
控制报文	4	源抑制
	5	路由重定向
查询报文	8/0	回送请求/应答
	13/14	时间戳请求/应答
	17/18	掩码请求/应答
	10/19	路由器请求/通告

在表 7-1 中，差错报告报文和控制报文为单向 ICMP 报文类型，表示路由器（或目的节点）发送到源节点的 ICMP 报文；查询报文为双向 ICMP 报文类型，发生在发送节点与目的节点（或路由器）之间，掩码请求/应答和路由器请求/通告不再使用。ICMP 报文数据区包含错误分组首部及该分组前 64 位数据（TCP/IP 协议规定，各协议都要把重要信息包含在这 64 位数据中），目的在于帮助源主机分析分组原因。虽然 ICMP 报文被封装在 IP 分组中，并作为 IP 分组的数据部分向外发送，但是并不能就此把 ICMP 协议看作一种高层协议。如前所述，ICMP 协议只是网络层协议的一部分，其辅助 IP 协议提供差错报告报文传输机制。

7.2.2 ICMP 差错报告报文语义及同步关系

ICMP 协议最基本的功能就是辅助 IP 协议提供差错报告报文传输机制。对于差错的处理方式，ICMP 协议没有严格的规定。事实上，当源主机收到 ICMP 差错报告报文后，还需要与应用程序联系起来，才能决定相应的差错处理方式。ICMP 差错报告报文都是采用路由器向源主机单向报告的模式，即当路由器发现 IP 分组出现差错后，调用 ICMP 协议，该 IP 分组的源主机发送 ICMP 差错报告报文告知其差错原因。同时，发生差错的 IP 分组被网络交换节点丢弃，不再向前转发。在 ICMP 差错报告报文中，有目的不可达报文、超时报文和参数出错报文等，其中，目的不可达报文依据情况可分为不同的子类型，如表 7-2 所示。

表 7-2 目的不可达报文子类型

类型	代码	含义	发送者
3	0	网络不可达，路由器找不到目的网络路由	路由器
	1	主机不可达，到达目的网络后，路由器发送 ARP 请求报文不成功	路由器
	2	协议不可达，IP 分组携带的数据属于高层协议，目的主机高层协议进程（传输层）没有运行	目的主机
	3	端口不可达（应用进程），目的主机应用进程没有运行	目的主机

1）目的不可达报文

路由器的主要功能是为 IP 分组选择路由并转发。当路由器从路由表中查不出与 IP 分组的目的 IP 地址对应的路由时，会发生目的不可达的错误，目的不可达 ICMP 报文的类型为 3，此时子类型代码为 0，表示网络不可达。这时，路由器要向源主机发送网络不可达的 ICMP 报文。目的不可达 ICMP 报文的类型为 3，并进一步细分成 13 种子类型，如网络不可达、主机不可达、协议不可达及端口不可达等，用代码来标识不同子类型，其他信息字段未用，全为 0。

2）超时报文

分组每经过一个路由器，其生存期（TTL）都要减 1，如果 TTL 递减为 0，则路由器会丢弃该分组，并向源主机发送报文类型为 11、代码为 0 的 ICMP 报文，报告该分组丢弃原因为超时。如果目的主机在对分组进行组装的过程中发生组装超时，则丢弃现已收到的各分片，并在第 1 个分片到达后向源主机节点发送报文类型为 11、代码为 1 的 ICMP 超时差错报告报文。Windows 系统下的 Tracert 命令及 Linux 系统下的 Traceroute 命令均是依据此 ICMP 差错报告报文来实现的。

3）参数出错报文

如果路由器或目的主机在对收到的 IP 分组进行首部校验处理时发现 IP 分组首部字段有错误，则将该 IP 分组丢弃，并向源主机发送报文类型为 12、代码为 0 的 ICMP 差错报告报文，并在 ICMP 差错报告报文的其他信息字段中，用 1 个字节作为指针来指出差错在 IP 分组首部字段中的位置（以字节为单位），数据字段存放该 IP 分组首部及该分组前 64 位数据。对于差错处理方式，ICMP 协议没有严格的规定。事实上，源主机 IP 层收到 ICMP 差错报告报文后，还需要上层协议的配合才能决定采取的差错处理方式，如图 7-2 所示。

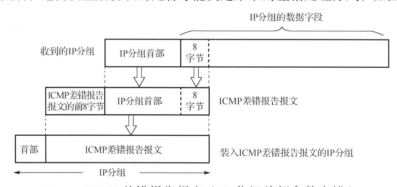

图 7-2 ICMP 差错报告报文（IP 分组首部参数出错）

注意事项：如果携带 ICMP 差错报告报文的 IP 分组在网络传输过程中出错，则不再产生 ICMP 差错报告报文；针对多个分片出错，只发送一个 ICMP 差错报告报文；对于组播 IP 报文出错，不产生 ICMP 差错报告报文；对于特殊 IP 地址的 IP 分组出错，如 127.x.y.z 或 0.0.0.0，不产生 ICMP 差错报告报文；在所有的 ICMP 差错报告报文中，只有 IP 分组首部参数出错报文包括数据字段，该字段包括源 IP 分组首部+源 IP 分组数据字段前 8 个字节（64 比特）；在图 7-2 中，IP 分组首部参数出错的 ICMP 差错报告报文的前 8 个字节数据提供了关于 UDP 和 TCP 端口号的信息，源主机根据协议号将差错情况报告给 TCP 或 UDP 协议，再通过端口号报告给应用进程。

7.2.3 ICMP 控制报文语义及同步关系

ICMP 控制报文主要包括源抑制报文和路由重定向报文，分别用于网络拥塞控制和路由重定向。

（1）源抑制报文：主要用于网络拥塞控制，目的是抑制源主机节点发送数据报文的速率。当网络交换节点由于缺乏缓存空间而无法接收新分组时，处理办法是载荷脱落机制，即按照一定的策略丢弃后续接收到的 IP 分组，并向源主机发送报文类型为 4、代码为 0 的源抑制报文。当源主机收到源抑制报文后，降低其分组发送速率，直到不再收到源抑制报文为止，然后源主机又逐渐增加分组发送速率，直到再次接收到源抑制报文为止。

拥塞控制的概念与流量控制有所不同。网络拥塞控制产生的直接原因是网络交换节点缓存空间不够，当路由器的数据报输入速率超过路由器的转发速度时，可能发生拥塞现象。拥塞控制带有全局性，因为拥塞可能影响到整个网络的数据传输，所以需要各节点共同参与协同解决。端到端流量控制产生的主要原因是接收终端节点传输层接收缓存不足，造成接收缓存输入速率大于向上层交付的速率，由于输入速率近似等于接收速率，接收速率可近似等于发送终端的发送速率，因此发送终端的发送速率大于向上层交付的速率，产生传输层接收缓存溢出现象。对于路由器和目的主机而言，理想的源抑制报文发送时间是当前输入分组数量占用的缓存区容量接近系统限制，这样可以避免因为输入的后续分组占用的缓存区容量超过系统限制而被丢弃，从而可以减少网络重发的分组量。

（2）重定向报文：提供了一种路由优化控制机制，使源主机能以动态方式寻找到最短路由，通常重定向报文只能在同一网段中的源主机与路由器之间使用。在图 7-3 中，假设 PC1 的默认路由器为 R1，PC1 向 PC2 发送 IP 分组，当路由器 R1 从处于同一网段的主机 PC1 收到一个需转发的 IP 分组时，路由器 R1 将检查自身的路由表信息，选定下一个路由器 R2 继续转发该分组；如果这时路由器 R1 确认路由器 R2 的一个接口和 PC1 处于同一网段时，路由器 R1 就向 PC1 发送重定向报文，通知 PC1 将分组直接发给路由器 R2 将会是一条较短的转发路径；在重定向报文的其他信息字段中，要填入重定向的路由器（如路由器 R2）的 IP 地址。

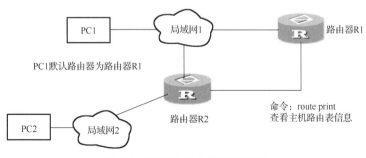

图 7-3 重定向报文应用示意图

7.2.4 ICMP 查询报文语义及同步关系

（1）回送请求/应答报文：主要用于测试网络目的节点（或路由器某个接口）的可达性。

当源节点向网络中某一特定的目的主机发送回送请求报文时，目的节点收到请求后必须使用回送应答报文来响应对方。ping 命令就是利用这种回送请求/应答报文来测试目的可达性的，具体应用如下。

① 命令：

```
Ipconfig /all;        //查看主机网络配置信息
```

② 命令：

```
ping 127.0.0.1;       //测试 TCP/IP 协议是否安装
```

③ 命令：

```
ping 本机 IP;         //测试本机网络配置信息正确性
```

④ 命令：

```
ping 网关 IP;         //测试本地主机到网关的可达性
```

（2）时间戳请求/应答报文：可用于估算源节点和目的节点间的分组往返时间。在该类型报文中使用了 3 个时间戳字段：初始时间戳字段是源节点发送时间戳请求报文的时间；接收时间戳字段是目的节点接收到时间戳请求报文的时间；发送时间戳字段是目的节点发送时间戳应答报文的时间。源节点首先发送时间戳请求报文，然后等待目的节点返回其应答报文，并根据这 3 个时间戳字段较为精确地估算两个节点间的报文往返时间。

（3）掩码请求/应答报文：主要用于主机获取所在网络的 IP 地址掩码信息。主机在发送掩码请求报文时，将 IP 分组首部中的源 IP 地址和目的 IP 地址字段的网络号部分设为 0。网络上的目的节点（通常为路由器）接收到该请求后，填写好网络的掩码信息并向源节点回送应答报文。

（4）路由器询问/通告报文：用于查找连接到源主机或路由器上的正常工作的其他路由器，并从中选择一个作为源主机（或路由器）的默认路由。

7.3 ping 命令实现分析

ping 命令用来测试两个主机之间的连通性。ping 命令使用了回送请求报文与回送应答报文。源节点向目的节点发送回送请求报文，目的节点接收到回送请求报文后给源节点返回一个回送应答报文；ping 命令是应用层直接使用网络层 ICMP 服务功能的例子，它没有通过运输层的 TCP 或 UDP 协议。ping 命令的输出信息主要包括发送报文数、接收报文数、丢失报文数，以及往返时间最小值、最大值和平均值。

```
#include <stdio.h>
#include <winsock2.h>
#pragma comment(lib, "ws2_32.lib")        /*WinSock 使用的库函数*/
/*宏定义 ICMP 报文类型*/
#define ICMP_TYPE_ECHO            8
#define ICMP_TYPE_ECHO_REPLY      0
```

```c
#define ICMP_MIN_LEN        8           /*ICMP 报文最小长度,只有首部*/
#define ICMP_DEF_COUNT      4           /*默认传输数据次数*/
#define ICMP_DEF_SIZE       32          /*默认数据长度*/
#define ICMP_DEF_TIMEOUT    1000        /*默认超时时间,毫秒*/
#define ICMP_MAX_SIZE       65500       /*最大数据长度*/

/*定义 IP 分组首部数据结构*/
struct ip_hdr
{
    unsigned char vers_len;             /*版本号和首部长度*/
    unsigned char tos;                  /*服务类型*/
    unsigned short total_len;           /*总长度*/
    unsigned short id;                  /*标识*/
    unsigned short frag;                /*标志和片偏移*/
    unsigned char ttl;                  /*TTL*/
    unsigned char proto;                /*协议类型*/
    unsigned short checksum;            /*校验和*/
    unsigned int sour;                  /*源 IP 地址*/
    unsigned int dest;                  /*目的 IP 地址*/
};

/*定义 ICMP 报文首部数据结构*/
struct icmp_hdr
{
    unsigned char type;                 /*类型*/
    unsigned char code;                 /*代码*/
    unsigned short checksum;            /*校验和*/
    unsigned short id;                  /*标识*/
    unsigned short seq;                 /*序号*/
    /*其他字段,不是标准 ICMP 首部,用于记录时间*/
    unsigned long timestamp;
};
//定义输出信息数据结构
struct icmp_user_opt
{
    unsigned int persist;               /*一直 Ping*/
    unsigned int count;                 /*发送 ECHO 请求的数量*/
    unsigned int size;                  /*发送数据的大小*/
    unsigned int timeout;               /*等待答复的超时时间*/
    char         *host;                 /*主机地址*/
    unsigned int send;                  /*发送数量*/
    unsigned int recv;                  /*接收数量*/
    unsigned int min_t;                 /*最短时间*/
    unsigned int max_t;                 /*最长时间*/
    unsigned int total_t;               /*总的累计时间*/
```

```c
};

/*随机数据*/
const char icmp_rand_data[] = "abcdefghigklmnopqrstuvwxyz0123456789"
                              "ABCDEFGHIJKLMNOPQRSTUVWXYZ";

struct icmp_user_opt user_opt_g = {
    0, ICMP_DEF_COUNT, ICMP_DEF_SIZE, ICMP_DEF_TIMEOUT, NULL,
    0, 0, 0xFFFF, 0
};

unsigned short ip_checksum(unsigned short *buf, int buf_len);

//构造 ICMP 数据
//参数说明：[IN, OUT] icmp_data, ICMP 缓存区；[IN] data_size, icmp_data 的长度
//[IN] sequence，序号
void icmp_make_data(char *icmp_data, int data_size, int sequence)
{
    struct icmp_hdr *icmp_hdr;
    char *data_buf;
    int data_len;
    int fill_count = sizeof(icmp_rand_data) / sizeof(icmp_rand_data[0]);

    /*填写 ICMP 报文数据*/
    data_buf = icmp_data + sizeof(struct icmp_hdr);
    data_len = data_size - sizeof(struct icmp_hdr);

    while (data_len > fill_count)
    {
        memcpy(data_buf, icmp_rand_data, fill_count);
        data_len -= fill_count;
    }

    if (data_len > 0)
        memcpy(data_buf, icmp_rand_data, data_len);

    /*填写 ICMP 报文首部*/
    icmp_hdr = (struct icmp_hdr *)icmp_data;

    icmp_hdr->type = ICMP_TYPE_ECHO;
    icmp_hdr->code = 0;
    icmp_hdr->id = (unsigned short)GetCurrentProcessId();
    icmp_hdr->checksum = 0;
    icmp_hdr->seq = sequence;
    icmp_hdr->timestamp = GetTickCount();
```

```c
    icmp_hdr->checksum = ip_checksum((unsigned short*)icmp_data, data_size);
}

//功能：解析接收到的数据
//参数说明：[IN] buf，数据缓存区；[IN] buf_len，buf 的长度；[IN] from，对方的地址
//返回值：成功返回 0，失败返回 -1
int icmp_parse_reply(char *buf, int buf_len,struct sockaddr_in *from)
{
    struct ip_hdr *ip_hdr;
    struct icmp_hdr *icmp_hdr;
    unsigned short hdr_len;
    int icmp_len;
    unsigned long trip_t;

    ip_hdr = (struct ip_hdr *)buf;
    hdr_len = (ip_hdr->vers_len & 0xf) << 2 ;  /*IP首部长度*/

    if (buf_len < hdr_len + ICMP_MIN_LEN)
    {
        printf("[Ping] Too few bytes from %s\n", inet_ntoa(from->sin_addr));
        return -1;
    }

    icmp_hdr = (struct icmp_hdr *)(buf + hdr_len);
    icmp_len = ntohs(ip_hdr->total_len) - hdr_len;

    /*检查校验和*/
    if (ip_checksum((unsigned short *)icmp_hdr, icmp_len))
    {
        printf("[Ping] icmp checksum error!\n");
        return -1;
    }

    /*检查 ICMP 报文类型*/
    if (icmp_hdr->type != ICMP_TYPE_ECHO_REPLY)
    {
        printf("[Ping] not echo reply : %d\n", icmp_hdr->type);
        return -1;
    }

    /*检查 ICMP 的标识*/
    if (icmp_hdr->id != (unsigned short)GetCurrentProcessId())
    {
        printf("[Ping] someone else's message!\n");
        return -1;
    }
```

```c
    /*输出应答信息*/
    trip_t = GetTickCount() - icmp_hdr->timestamp;
    buf_len = ntohs(ip_hdr->total_len) - hdr_len - ICMP_MIN_LEN;
    printf("%d bytes from %s:", buf_len, inet_ntoa(from->sin_addr));
    printf(" icmp_seq = %d  time: %d ms\n",icmp_hdr->seq, trip_t);

    user_opt_g.recv++;
    user_opt_g.total_t += trip_t;

    /*记录返回时间*/
    if (user_opt_g.min_t > trip_t)
        user_opt_g.min_t = trip_t;

    if (user_opt_g.max_t < trip_t)
        user_opt_g.max_t = trip_t;

    return 0;
}

//功能：接收数据，处理应答
//参数说明：[IN] icmp_soc，套接字描述符
//返回值：成功返回 0，失败返回 -1
int icmp_process_reply(SOCKET icmp_soc)
{
    struct sockaddr_in from_addr;
    int result, data_size = user_opt_g.size;
    int from_len = sizeof(from_addr);
    char *recv_buf;

    data_size += sizeof(struct ip_hdr) + sizeof(struct icmp_hdr);
    recv_buf = malloc(data_size);

    /*接收数据*/
    result = recvfrom(icmp_soc, recv_buf, data_size, 0,
                (struct sockaddr*)&from_addr, &from_len);
    if (result == SOCKET_ERROR)
    {
        if (WSAGetLastError() == WSAETIMEDOUT)
            printf("timed out\n");
        else
            printf("[PING] recvfrom_ failed: %d\n", WSAGetLastError());

        return -1;
    }
```

```
        result = icmp_parse_reply(recv_buf, result, &from_addr);
        free(recv_buf);

        return result;
}

//功能: 显示ECHO的帮助信息
//参数说明: [IN] prog_name, 程序名
//返回值: 无返回值

void icmp_help(char *prog_name)
{
        char *file_name;

        file_name = strrchr(prog_name, '\\');
        if (file_name != NULL)
            file_name++;
        else
            file_name = prog_name;

        /*显示帮助信息*/
        printf(" usage:     %s host_address [-t] [-n count] [-l size] "
            "[-w timeout]\n", file_name);
        printf(" -t         Ping the host until stopped.\n");
        printf(" -n count   the count to send ECHO\n");
        printf(" -l size    the size to send data\n");
        printf(" -w timeout timeout to wait the reply\n");
        exit(1);
}

//功能: 解析命令行选项, 保存到全局变量中
//参数说明: [IN] argc, 参数的个数; [IN] argv, 字符串指针数组
//返回值: 无返回值
void icmp_parse_param(int argc, char **argv)
{
        int i;

        for(i = 1; i < argc; i++)
        {
            if ((argv[i][0] != '-') && (argv[i][0] != '/'))
            {
                /*处理主机名*/
                if (user_opt_g.host)
                    icmp_help(argv[0]);
                else
                {
```

```c
                user_opt_g.host = argv[i];
                continue;
            }
        }

        switch (tolower(argv[i][1]))
        {
        case 't':       /*持续 Ping*/
            user_opt_g.persist = 1;
            break;

        case 'n':       /*发送请求的数量*/
            i++;
            user_opt_g.count = atoi(argv[i]);
            break;

        case 'l':       /*发送数据的大小*/
            i++;
            user_opt_g.size = atoi(argv[i]);
            if (user_opt_g.size > ICMP_MAX_SIZE)
                user_opt_g.size = ICMP_MAX_SIZE;
            break;

        case 'w':       /*等待接收的超时时间*/
            i++;
            user_opt_g.timeout = atoi(argv[i]);
            break;

        default:
            icmp_help(argv[0]);
            break;
        }
    }
}


int main(int argc, char **argv)
{
    WSADATA wsaData;
    SOCKET icmp_soc;
    struct sockaddr_in dest_addr;
    struct hostent *host_ent = NULL;

    int result, data_size, send_len;
    unsigned int i, timeout, lost;
    char *icmp_data;
```

```c
unsigned int ip_addr = 0;
unsigned short seq_no = 0;

if (argc < 2)
   icmp_help(argv[0]);

icmp_parse_param(argc, argv);
WSAStartup(MAKEWORD(2,0),&wsaData);

/*解析主机地址*/
user_opt_g.host = "10.12.1.188";
ip_addr = inet_addr(user_opt_g.host);
if (ip_addr == INADDR_NONE)
{
   host_ent = gethostbyname(user_opt_g.host);
   if (!host_ent)
   {
       printf("[PING] Fail to resolve %s\n", user_opt_g.host);
       return -1;
   }

   memcpy(&ip_addr, host_ent->h_addr_list[0], host_ent->h_length);
}

icmp_soc = socket(AF_INET, SOCK_RAW, IPPROTO_ICMP);
if (icmp_soc == INVALID_SOCKET)
{
   printf("[PING] socket() failed: %d\n", WSAGetLastError());
   return -1;
}

/*设置选项，接收和发送的超时时间*/
timeout = user_opt_g.timeout;
result = setsockopt(icmp_soc, SOL_SOCKET, SO_RCVTIMEO,
               (char*)&timeout, sizeof(timeout));

timeout = 1000;
result = setsockopt(icmp_soc, SOL_SOCKET, SO_SNDTIMEO,
               (char*)&timeout, sizeof(timeout));

memset(&dest_addr,0,sizeof(dest_addr));
dest_addr.sin_family = AF_INET;
dest_addr.sin_addr.s_addr = ip_addr;

data_size = user_opt_g.size + sizeof(struct icmp_hdr) - sizeof(long);
icmp_data = malloc(data_size);
```

```c
    if (host_ent)
        printf("Ping %s [%s] with %d bytes data\n", user_opt_g.host,
            inet_ntoa(dest_addr.sin_addr), user_opt_g.size);
    else
        printf("Ping [%s] with %d bytes data\n", inet_ntoa(dest_addr.sin_addr),
            user_opt_g.size);

    /*发送请求并接收应答报文*/
    for (i = 0; i < user_opt_g.count; i++)
    {
        icmp_make_data(icmp_data, data_size, seq_no++);

        send_len = sendto(icmp_soc, icmp_data, data_size, 0,
                        (struct sockaddr*)&dest_addr, sizeof(dest_addr));
        if (send_len == SOCKET_ERROR)
        {
            if (WSAGetLastError() == WSAETIMEDOUT)
            {
                printf("[PING] sendto is timeout\n");
                continue;
            }

            printf("[PING] sendto failed: %d\n", WSAGetLastError());
            break;
        }

        user_opt_g.send++;
        result = icmp_process_reply(icmp_soc);

        user_opt_g.persist ? i-- : i;       /*持续 Ping*/
        Sleep(1000);                        /*延迟 1s*/
    }

    lost = user_opt_g.send - user_opt_g.recv;

    /*打印统计数据*/
    printf("\nStatistic :\n");
    printf("    Packet : sent = %d, recv = %d, lost = %d (%3.f%% lost)\n",
        user_opt_g.send, user_opt_g.recv, lost, (float)lost*100/user_opt_g.send);

    if (user_opt_g.recv > 0)
    {
        printf("Roundtrip time (ms)\n");
        printf("    min = %d ms, max = %d ms, avg = %d ms\n", user_opt_g.min_t,
            user_opt_g.max_t, user_opt_g.total_t / user_opt_g.recv);
```

```
    }

    free(icmp_data);
    closesocket(icmp_soc);
    WSACleanup();

    return 0;
}

//函数功能：计算校验和
//参数说明：[IN] buf,数据缓存区； [IN] buf_len,buf 的字节长度
//返回值：校验和
unsigned short ip_checksum(unsigned short *buf, int buf_len)
{
    unsigned long checksum = 0;

    while (buf_len > 1)
    {
        checksum += *buf++;
        buf_len -= sizeof(unsigned short);
    }

    if (buf_len)
    {
        checksum += *(unsigned char *)buf;
    }
    checksum = (checksum >> 16) + (checksum & 0xffff);
    checksum += (checksum >> 16);
    return (unsigned short)(~checksum);
}
```

7.4 Tracert 命令设计与实现

Linux 系统下的 Traceroute 命令是 Van jacobson 在 1988 年 12 月设计并实现的，主要用于探测从源节点到目的节点的参考路径信息。用户主机不论采用发送不同 TTL 的 ICMP 报文，还是 UDP 用户数据报，均以 IP 分组形式在网络上进行传输。由于网络负载可能发生变换，因此不同 TTL 的 IP 分组在传输时可能选用不同的路径，从这个意义上说，Tracert 命令获得的路径信息仅具有参考价值，不一定完全正确，但一般情况下这个可能性比较低，如图 7-4 所示。如果发送 ICMP ECHO 请求报文进行探测，则网络设备由于管理员设置了屏蔽 ICMP 报文功能，因此网络设备不会向源主机发送 ICMP 超时差错报告报文和 ICMP ECHO 应答报文，此时用户主机无法得到该路由器 IP 地址，所以在屏幕上显示"*"。在此情况下，如果用户主机收到 ICMP 超时差错报告报文，则说明没有到达目的方；如果用户主机收到 ICMP ECHO 应答报文，则说明已经到达目的方。

```
C:\Documents and Settings\XXR>tracert mail.sina.com.cn

Tracing route to mail.sina.com.cn [202.108.43.230]
over a maximum of 30 hops:

  1    24 ms    24 ms    23 ms  222.95.172.1
  2    23 ms    24 ms    22 ms  221.231.204.129
  3    23 ms    22 ms    23 ms  221.231.206.9
  4    24 ms    23 ms    24 ms  202.97.27.37
  5    22 ms    23 ms    24 ms  202.97.41.226
  6    28 ms    28 ms    28 ms  202.97.35.25
  7    50 ms    50 ms    51 ms  202.97.36.86
  8   308 ms   311 ms   310 ms  219.158.32.1
  9   307 ms   305 ms   305 ms  219.158.13.17
 10   164 ms   164 ms   165 ms  202.96.12.154
 11   322 ms   320 ms  2988 ms  61.135.148.50
 12   321 ms   322 ms   320 ms  freemail43-230.sina.com [202.108.43.230]

Trace complete.
```

图 7-4　Tracert 命令应用示意图

如果源节点通过发送 UDP 用户数据报进行探测，则每次将 UDP 探测报文端口号加 1，由于 UDP 用户数据报首部会被封装在 ICMP 超时差错报告报文的数据部分中，因此可以匹配接收的 ICMP 超时差错报告报文，同时可以解决 ICMP 报文受管理员屏蔽的限制。在此情况下，如果源节点收到 ICMP 超时差错报告报文，则说明没有到达目的节点；由于目的节点端口号一般比较大，目的节点不经常使用，因此当该 UDP 用户数据报到达目的节点时，如果目的节点端口号对应进程没有启动，则向源主机发送目的端口不可达 ICMP 差错报告报文，说明探测报文已到达目的节点。源节点向目的节点发送一连串的 UDP 用户数据报（实际上在网络层是以不同 IP 分组传输的），分组中封装的是无法交付的 UDP 用户数据报。源节点首先向目的节点发送第 1 个 UDP 用户数据报，并且 TTL=1；当该分组到达路由器 R1 时，路由器 R1 先接收，然后 TTL–1=0，不再转发，直接丢弃，并向源节点发送一个 ICMP 超时差错报告报文，源节点获得路由器 R1 的入口 IP 地址。源节点向目的节点发送第 2 个 UDP 用户数据报，并且 TTL=2；当该分组到达路由器 R2 时，路由器 R2 先接收，然后 TTL-1=0，不再转发，直接丢弃，并向源节点发送一个 ICMP 超时差错报告报文，源节点记录下路由器 R2 的入口 IP 地址。依次类推，当最后一个分组到达目的节点时，目的节点不转发，则 IP 分组中封装的 UDP 用户数据报无法交付，这时目的节点向源节点发送一个"目的端口不可达"差错报告报文，表明分组已经到达目的节点。

在 Windows 环境下，一般使用 Tracert 命令，其在一定意义上可以跟踪一个分组从源节点到目的节点的参考路径。在 Win XP 环境下，通过 ICMP 探测报文和 UDP 探测报文实现 Tracert 命令均没有问题；在 Windows vista 环境下，使用 ICMP 探测报文实现 Tracert 命令没有问题；当使用 UDP 探测报文时，创建的 ICMP 原始套接字上无法收到从网络设备返回的 ICMP 超时差错报告报文和目的不可达报文，但 Wireshark 抓包工具可以捕获到 ICMP 超时差错报告报文和目的不可达报文。如果在 Windows vista 环境下使用同一个 UDP 套接字发送和接收，recvfrom()函数会返回 WSAECONNRSET 和 WSAENETRESET 错误信息。WSAECONNRSET 表示在 UDP 套接字上发生 TTL 超时；WSAENETRESET 表示在 UDP 套接字上产生目的不可达信息，但无法使用 recvfrom()函数直接获得对方的 IP 地址。

Tracert 命令实现源代码如下。

```c
#include "Winsock2.h"
#include "stdio.h"
#include "Ws2tcpip.h"

#pragma comment(lib,"ws2_32.lib")

//定义UDP伪首部数据结构,在计算UDP首部校验和时用
Typedef struct tagPseUDPHeader
{
    Unsigned int srcAddress;           //源IP地址号
    Unsigned int dstAddress;           //目的IP地址号
    Unsigned char reserved;            //保留
    unsigned char protocolType;        //协议类型
    Unsigned short UDPTotalLen;        //UDP用户数据报总长度(首部+数据)
}

//定义UDP用户数据报包首部数据结构
typedef struct   tagUDPHeader
{
    unsigned short src_port;           //16位源端口号
    unsigned short dst_port;           //16位目的端口号
    unsigned short udp_length;         //16位总长度(首部+数据)
    unsigned short udp_checksum;       //16位简单校验和
}

 //定义ICMP报文首部数据结构
Struct tagICMPHeader
{
   Unsigned char type
Unsigned char code;
Unsigned short checksum;
Unsigned short id;
Unsigned short seq;
}
//定义IP分组首部数据结构
typedef struct   tagIPHeader
{
    unsigned char ip_verlen;           //4位首部长度+4位版本号
    unsigned char ip_tos;              //8位服务类型
    unsigned short ip_totallength;     //16位总长度(字节)
    unsigned short ip_id;              //16位标识
    unsigned short ip_offset;          //3位标志+片偏移
    unsigned short ip_ttl;             //8位TTL
    unsigned short ip_protocol;        //8位协议类型(TCP、UDP或其他)
    unsigned short ip_checksum;        //16位IP首部简单校验和
    unsigned long ip_srcaddr;          //32位源IP地址
```

```c
    unsigned long ip_destaddr;           //32位目的IP地址
}

Struct tagTracertControl
{
    Unsigned int max_hop;                //搜索目标的最大跳数
    Unsigned int timeout;                //等待应答的超时时间：毫秒
    Unsigned int use_udp;                //发送UDP探测报文
    Unsigned int timestamp;              //发送时间戳
    Char         *hostname;              //主机名称
    Unsigned short sport;                //源端口号，网络字节序列
}

//ICMP报文类型
#define ICMP_ECHO_REQUEST 8
#define ICMP_ECHO_REPLY   0
#define ICMP_DEST_UNREACHED 3
#define ICMP_TTL_EXCEEDED 11

//ICMP消息代码（子类型）
#define ICMP_UNREACH_PORT  3
#define ICMP_EXCEED_TRANS  0

#define ICMP_HDR_SIZE       8            //ICMP报文首部最小值

#define TRACERT_MAX_TIMES   3            //消息发送的次数
#define TRACERT_MAX_HOP     30           //最大跳数
#define TRACERT_DEF_TIMEOUT 5000         //接收的超时时间，毫秒
#define TRACERT_PKT_SIZE    32           //发送分组的大小
#define TRACERT_PKT_MAX     256          //接收的最大缓存区
#define TRACERT_DEST_PORT 32700          //UDP探测报文目的端口号


//宏定义函数返回值
#define TRACERT_DONE -1;
#define TRACERT_EXCEEDED -2
#define TRACERT_TIMEOUT -3
#define TRACERT_ERROR -4

//命令帮助信息
Void tracert_help(char *program_name)
{
    Char *file_name;
    Filename =strrchar(prog_name,"\\");
    If(filename != NULL)
      File_name++;
```

```
    Else
        File_name = programe_name;

    Printf("Usage help: %s host name [-h hops] [-w timeout] [-u udp]",filename);
    Print("\n please use the option:\n");
    Printf("-h the maximum hops to the destination \n");
        Printf("-w timeout milliseconds for each reply \n");
    Printf("-u use udp packet as the prob \n");
    Printf("\n example:%s www.china.com -h 10 -w 1000 \n \n", filename);
}

//对用户输入的参数进行解析

Void parse_param(int argc, char *argv[])
{
    Int I;
If (argc < 2)
{
    Tracert_help(argv[0]);
}
For (int I =1; I <argc; i++)
{
    If ((argv[i][0] != '-') && (argv[i][0] != '/') )
    {
        If(tracert_ctrl_param.host)
            Tracert_help ([argv[0]);
        Else
        {
            Tracert_ctrl_param.host =argv[i];
            Continue;
        }
    }

    Switch(tolower(argv[i][1]))
    {
        Case:'h'
            I++;
            Tracert_ctrl_param.max_hop = atoi (argv[i]);
            Break;
        Case:'w
            I++;
            Tracert_ctrl_param.timeout=atoi (argv[i]);
            Break;
        Case:'u'
            Tracert_ctrl_param.udp = 1;
            Break;
```

```
        Default:
            Tracert_help(argv[0]);
        break
    }
}

Return;
}

//根据用户输入的域名获得IP地址
Unsigned long tracert_resolve_ip_addr(char *hostname)
{
    Unsigned long ipaddr;
Struct hostent *host_ent = NULL;

Ipaddr = inet_addr(hostname);
If (ipaddr != IANDDR_NONE)
{
    Return ipaddr;
}

Host_ent = gethostbyname(host);
If(host_ent)
    Memcpy(&ipaddr, host_ent->h_addr_list[0],host_ent->h_length);

Return ipaddr;
}

//发送ICMP ECHO请求报文
Int tracert_icmp_send(SOCKET socket, struct sockaddr *dest_addr, int ttl,
        short seq)
{
    Int I, data_len = 0,result =0,msg_len = TRACERT_PKT_SIZE;
    Char icmp_data[TRACER_PKT_SIZE],ptrChar;
    struct tagICMPHeader *ptrICMPHeader;
    //构造ICMP ECHO请求报文
    ptrICMPHeader = (struct tagICMPHeader *)icmp_data;
    ptrICMPHeader -> type = ICMP_ECHO;
     ptrICMPHeader1 -> code = 0;
    ptrICMPHeader -> checksum = 0;
    ptrICMPHeader -> id = (unsigned short) GetCurrentProcessId();
    ptrICMPHeader -> seq = seq;
    ptrChar = icmp_data + sizeof (struct tagICMPHeader);
    data_len = msg_len - sizeof(struct tagICMPHeader);

//初始化ICMP ECHO报文数据
```

```
For(I =0; i<data; i++)
{
   ptrChar[i] = 'A' + I;
}

   ptrICMPHeader -> checksum = ip_checksum((unsigned short *)icmp_data,msg_len);

   //设置 TTL 并发送数据
   Result = setsockopt(sock,IPPROTO_IP,IP_TTL,
          (char *)&ttl,sizeof(int));
   Tracert_ctl_param.timestamp = GetTickCount();
   Result = sendto(socket,icmp_data,msg_len,0,dest_addr,sizeof(*dest_addr));
   Return result;
}


//发送 UDP 探测报文
Int tracert_udp_send(SOCKET socket, struct sockaddr *dest, int ttl short seq)
{
   Int I, result = 0;
   Char data[TRACERT_PKT_SIZE];
   Struct sockaddr_in *dest_addr = (struct sockaddr_in *)dest;

   //初始化 UDP 用户数据报数据
   For (I =0; i< TRACERT_PKT_SIZE, i++)
     Data[i] = 'A' + I;
   //设置 TTL 初值
   Result = setsockopt(socket,IPPROTO_IP,IP_TTL,(char *)&ttl, sizeof(int));
   }
   Tracert_ctrl_param.timestamp = GetTickCount();

   Dest_addr ->sin_port htons((unsigned short)(TRACERT_DEST_PORT + seq));

   Result = sendto(socket, data, TRACERT_PKT_SIZE,0,dest, sizeof(*dest));
   Return result;
   }

   //对接收到的应答报文进行解析
Int tracert_parse_reply(char *buf, int buf_len, short seq)
{
   Struct tagIPHeader *ptrIPHeader,*ptrIPHeader1;
   Struct tagICMPHeader *ptrICMPHeader, *ptrICMPHeader1;
   Struct tatUDPHeader *ptrUDPHeader;
   Unsigned short iph_len,iph_len1;
   Unsigned short icmph_len,min_len;
    Unsigned short proc_id = (unsigned short) GetCurrentProcessId();
```

```
    Unsigned short dport = htons((unsigned short)(TRACERT_DEST_PORT + seq));
Char *proto_data;
Int imcp_type =0;

ptrIPHeader = (struct tagIPHeader *)buf;
iph_len = (ptrIPHeader -> vers_len & 0x0f) << 2;   //计算IP分组首部长度

if(buf_len < iph_len +ICMP_HDR_SIZE)
{
   Printf("too few bytes received!\n");
   Return -1
}

   Imcp_len = ntohs(ptrIPHeader -> total_len) - iph_len;
   ptrICMPHeader = (struct tagICMPHeader *)(buf + iph_len);
   icmp_type = ptrICMPHeader -> type;

//获取ICMP报文数据
ptrIPHeader1 = (struct tagIPHeader *)(buf + iph_len + ICMP_HDR_SIZE);
iph_len2= (ptrIPHeader1 -> vers_len &0x0f)<<2    //计算ICMP报文内IP首部长度
proto_data = buf + iph_len +ICMP_HDR_SIZE +iph_len1;

ptrICMPHeader1 = (struct tagICMPHeader *) proto_data;
ptrUDPHeader = (struct tagUDPHeader *) proto_data;

min_len = ICMP_HDR_SIZE +iph_len1+ICMP_HDR_SIZE;
if(icmp_len < min_len)
  {
    Printf(" ICMP too few bytes received!\n");
    Return -1
}
Switch(icmp_type)
{
   Case ICMP_ECHO_REPLY: //ICMP ECHO应答报文，说明ICMP请求报文已经到达目的方
      If(ptrICMPHeader -> id == proc_id)         //检查ICMP报文的标识字段
         Return TRACERT_DONE;
      Break;
   case ICMP_DEST_UNREACH:           //目的不可达报文，UDP探测到目的方
      If(ptrIPHeader1 -> proto == IPPROTO_UDP && ptrUDPHeader -> dport == dport
            && ptrUDPHeader -> sport == tracert_ctrl_param.sport)
      {
          If(ptrICMPHeader -> code == ICMP_UNREACH_PORT)
             Return TRACERT_DONE;
      }
      Break;
   Case ICMP_TTL_EXCEEDED:   //收到ICMP超时差错报告报文
```

```
    If(ptrICMPHeader -> code != ICMP_EXCEED_TRANS)
       Break;
    If(tracert_ctrl_param.use_udp)
    {
       If(ptrIPHeader1 -> proto == IPPROTO_UDP  && ptrUDPHeader ->dport ==
          dport && ptrUDPHeader ->sport == tracert_ctrl_param.sport)
       {
           Return TRACERT_EXCEEDED;
       }
    }
    Else if (ptrICMPHeader -> id == proc_id)
       Return TRACERT_EXCEEDED'
    Break;

    Default:
       Break;
    }

    Return icmp_type;
}

//接收UDP用户数据报并对ICMP报文首部进行解析
Int tracert_handle_reply(SOCKET icmp_sock, short seq,
        Struct sockaddr_in *from, long tracertip)
{
    Int result;
    Char recv_buf[TRACERT_PKT_MAX];
    Int from_len = sizeof(*from);

    While(true)
    {
       Result = recvfrom(icmp_sock, recv_buf, TRACERT_PKT_MAX, 0
          (struct sockaddr *) from ,&from_len);
       If(result == SOCKET_ERROR)
       {
           If(WSAGetLastError() == WSAETIMEDOUT))
           {
               Return TRACERT_TIMEOUT;
           }
           Printf("recvfrom () error:%d\n", WSAGetLastError());
           Return TRACERT_ERROR;
       }

       *tracertip = GetTickCount() - tracert_ctrl_param.timestamp;

       //解析ICMP报文
```

```c
        Result = tracert_parse_reply(recv_buf,result, seq);
        If(result <0)
            Return result;
    }
    Return result;


}
int main(int argc, char *argv[])
{
    WSADATA wsa_data;
    SOCKET  icmp_sock,udp_sock;
    Int I,result, max_hop,ttl = 1;
    Int done =0

    Int local_len =sizeif(struct sockaddr_in);

    Short seq = 0;
    Struct sockaddr_in dest_addr, from_addr,local_addr;
    Long trtip_tm = 0;

    Tracert_parse_param(argc, argv);

    WSAStartup(MAKEWORD(2,0), &wsa_data);

    //解析主机IP地址
    Memset(&dest_addr,0,sizeof(dest_addr));
    Dest_addr.sin_family = AF_INET;
    Dest_addr.sin_addr.s_addr = tracert_resolve_addr(tracert_ctrl_param.hostname);
    If(dest_addr.sin_addr.s_addr == INADDR_NONE)
    {
        Printf("unable to resolve dest host: %s\n", tracert_ctrl_param.hostname);
        Return -1;
    }
    //创建ICMP套接字
    Icmp_sock = socket(AF_INET,SOCK_RAW, IPPROTO_ICMP);

    //创建UDP套接字
    If(tracert_ctrl_param.use_udp)
    {
        //绑定ICMP套接字的本地地址
        Memset(&local_addr,0, sizeof(local_addr));
        Local_addr.sin_family = AF_INET;
        Loacal_addr.sin_addr.s_addr = INADDR_ANY;
        Bind(icmp_socket,(struct sockaddr *)local_addr,sizeof(local_addr));

        //绑定UDP套接字
```

```
   Udp_socket = socket(AF_INET, SOCK_DGRAM, 0);
   Bind(udp_socket,(struct sockaddr *)&local_addr, sizeof(local_addr));
   Getsockname(udp_socket,(struct sockaddr *)local_addr,&local_addr);

   Tracert_ctrl_param.sport = local_addr.sin_port;
}

//设置接收ICMP应答报文的最大等待时间
Result = setsockopt(icmp_sock, SOL_SOCKET, SO_RECVTIMEO,
         (char *)&tracert_ctrl_param.timeout, sizeof(int));
If(result == SOCKET_ERROR)
{
   Printf("setsocketoption error: %d\n", WSAGetLastError());
   Return -1
}
Max_hop = tracert_ctrl_param.max_hop;
Printf("\n tracing route to %s over a maximum of %d hops\n",
Tracert_ctrl_param.host, max_hops);

For (ttl =1; ttl < max_hop && !done; ttl++)
{
   Printf("\n%2d",ttl);
   For (I =0; i<TRACERT_MAX_TIMES, i++)
   {
      //发送UDP用户数据报或ICMP报文
      If(tracert_ctrl_param.use_udp)
      {
         Result = tracert_udp_send(udp_socket, dest_addr,ttl, ++seq);
      }
      Else
      {
         Result = tracert_icmp_send(icmp_socket,dest_addr,ttl, ++seq);
      }

      If (result == SOCKET_ERROR)
      {
         Printf("tracert send error: %d\n",WSAGetLastError());
         Go to end;
      }

      //接收ICMP报文并解析
      Result = tracert_handle_reply(icmp_socket, seq,&from_addr,&trip_tm);
      Switch(result)
      {
         Case TRCERT_TIMEOUT:
            Printf(" *  ");
```

```
                Break;
        Case TRACERT_EXCEEDED:
        Case TRCERT_DONE:
            Printf("%6d ms",trip_tm);
            Result = TRACERT_DONE? Done++:done;

            If(I == (TRACERT_MAX_TIMES -1))   //最后一次，输出地址
              Printf("  %s  ", inet_ntoa(from.sin_addr));
            Break;
        Case TRACERT_ERROR:
            Go to end;
            Break;
        Default:
            Printf("ICMP code %d",result);
            Break;
       }
    Sleep(1000);

      }
   }
}
End:
  Printf("\n\n tracert complete.\n");
  Closesocket(icmp_socket);
  If(tracert_ctrl_param.use_udp)
      Closesocket(udp_socket);

  WSACleanup();

Return 0;
}
```

7.5 ICMP 协议发送 ICMP ECHO 请求报文

ICMP 协议 ICMP ECHO 请求报文发送源代码如下。

```
#include "Winsock2.h"
#include "stdio.h"
#include "Ws2tcpip.h"

#define ICMP_TYPE_ECHO          8
#define ICMP_TYPE_ECHO_REPLY    0
#define ICMP_MIN_LEN            8       /*ICMP 报文最小长度，只有首部*/
#define ICMP_DEF_COUNT          4       /*默认数据次数*/
#define ICMP_DEF_SIZE           32      /*默认数据长度*/
#define ICMP_DEF_TIMEOUT        1000    /*默认超时时间，毫秒*/
```

```c
#define ICMP_MAX_SIZE          65500         /*最大数据长度*/

#pragma comment(lib,"ws2_32.lib")

//定义ICMP报文首部
typedef struct   tagICMPHeader
{
    unsigned char type;                      /*类型*/
    unsigned char code;                      /*代码*/
    unsigned short checksum;                 /*校验和*/
    unsigned short id;                       /*标识*/
    unsigned short seq;                      /*序号*/
}ICMPHeader, *ptrICMPHeader;

//定义IP分组首部数据结构
typedef struct   tagIPHeader
{
    unsigned char ip_verlen;                 //4位首部长度+4位版本号
    unsigned char ip_tos;                    //8位服务类型
    unsigned short ip_totallength;           //16位总长度(字节)
    unsigned short ip_id;                    //16位标识
    unsigned short ip_offset;                //3位标志+片偏移
    unsigned short ip_ttl;                   //8位TTL
    unsigned short ip_protocol;              //8位协议类型(TCP、UDP或其他)
    unsigned short ip_checksum;              //16位IP分组首部简单校验和
    unsigned long ip_srcaddr;                //32位源IP地址
    unsigned long ip_destaddr;               //32位目的IP地址
}IPHeader, *ptrIPHeader;

int main()
{
    //初始化要发送的数据
    const int BUFFER_SIZE = 1000;
    char icmp_buffer[BUFFER_SIZE];
    Unsigned short checkBuffer[65535];
    const char *icmpData = "create ICMP echo request packet and send!";

    //初始化Windows Socket DLL
    WSADATA wsd;
    if (WSAStartup(MAKEWORD(2, 2), &wsd) != 0)
    {
        printf("WSAStartup() failed: %d ", GetLastError());
        return -1;
    }
    //创建Raw Socket
    SOCKET rawSocket = WSASocket(AF_INET, SOCK_RAW, IPPROTO_IP, NULL, 0,
```

```
                        WSA_FLAG_OVERLAPPED);
if (rawSocket == INVALID_SOCKET)
{
    printf("WSASocket() failed: %d ", WSAGetLastError());
    return -1;
}

//设置发送超时时间
Int timeout = 1500;
Setsockopt(RawSocket, SOL_SOCKET, SO_SNDTIMEO, (char *)&timeout, sizeof
            (timeout));

//设置首部控制选项，使用 IP_HDRINCL
DWORD bOption  = TRUE;
int retResult= setsockopt(s, IPPROTO_IP, IP_HDRINCL, (char*) &bOption,
                sizeof(bOption));
if (retResult == SOCKET_ERROR)
{
    printf("setsockopt(IP_HDRINCL) failed: %d ", WSAGetLastError());
    closesocket(rawSocket);
    WSACleanup();
    return -1;
}
//填充IP分组首部各字段
IPHeader = {0};
IPHeader.ip_version = (0x04<<4|sizeof(IPHeader)/4);
IPHeader.ip_tos = 0;
IPHeader.ip_totallength = sizeif(IPHeader) + sizeof (ICMPHeader) +
                        sizeof(icmp_data);
IPHeader.ip_id = 0;
IPHeader.ip_ttl = 64;
IPHeader.ip_protocol=IPPROTO_ICMP;
IPHeader.ip_checksum = 0
IPHeader.ip_srcaddr = inet_addr("192.168.0.1");
IPHeader.ip_dstaddre = inet_addr("192.168.0.2");

//初始化ICMP ECHO REQUEST报文首部
ICMPHeader = {0};
ICMPHeader.type = ICMP_TYPE_ECHO;
ICMPHeader.code = 0;
ICMPHeader.checksum = 0;
ICMPHeader.id = (unsigned short)GetCurrentProcessId();;
ICMPHeader.seq = 0;

//计算ICMP ECHO REQUEST报文首部简单校验和
Memset(checkBuffer, 0, 65535);
```

```
    Memcpy(checkBuffer, &ICMPHeader,sizeof(ICMPHeader));
    Memcpu(checkBuffer + sizeif(ICMPHeader), & icmpData,sizeof(icmpDara));
    ICMPHeader .checksum = checksum(checkBuffer, sizeif(ICMPHeader) +
    + sizeof(icmpData));

    //构建待发送的 ICMP ECHO REQUEST 报文
    Memset(icmp_buffer, 0, 1000);
    Memcpy(icmp_buffer, &IPHeader, sizeof(IPHeader));
    Memcpy(icmp_buffer + sizeof(IPHeader),&ICMPHeader, sizeof(ICMPHeader));
    Memcpy(icmp_buffer +sizeOf(IPHeader) +sizeof(ICMPHeader),icmpData,sizeof(icmpData));
    //填充 Socketaddr_in 数据结构
    sockaddr_in  remote_addr;
    remote_addr.sin_family = AF_INET;
    remote_addr.sin_port = htons(8000);
    remote_addr.sin_addr.s_addr = inet_addr("192.168.0.2");
    //发送 UDP 用户数据报
    int iSendSize = size0f(IPHeader) +sizeof(ICMPHeader) + sizeof(icmpData);
    int  retResult = sendto(rawSocket, buffer, iSendSize, 0,
         (struct sockaddr *) &remote_addr, sizeof(remote_addr));
    if (retResult == SOCKET_ERROR)
       printf("sendto() failed: %d ", WSAGetLastError());
    else
       printf("sent %d bytes ", retResult);

    //关闭 Socket,释放资源
    closesocket(rawSocket);
    WSACleanup();
    return   true;
}
```

7.6 ICMP 协议接收 ICMP ECHO 请求报文

ICMP 协议 ICMP ECHO 请求报文接收源代码如下。

```
#include "Winsock2.h"
#include "stdio.h"
#include "Ws2tcpip.h"

#define ICMP_TYPE_ECHO         8      /*ICMP EACHO 请求报文类型*/
#define ICMP_TYPE_ECHO_REPLY   0      /*ICMP EACHO 应答报文类型*/
#define ICMP_MIN_LEN           8      /*ICMP 报文最小长度,只有首部*/
#define ICMP_DEF_SIZE          32     /*默认数据长度*/
#define ICMP_MAX_SIZE          1024   /*最大数据长度*/
#pragma comment(lib,"ws2_32.lib")

//定义 ICMP 报文首部
```

```c
typedef struct    tagICMPHeader
{
    unsigned char type;                 /*类型*/
    unsigned char code;                 /*代码*/
    unsigned short icmpheadercheck;     /*校验和*/
    unsigned short id;                  /*标识*/
    unsigned short seq;                 /*序号*/
}ICMPHeader, *ptrICMPHeader;

//定义 IP 分组首部数据结构
typedef struct    tagIPHeader
{
    Union
    {
        Unsigned  char version;         //一个字节前 4 比特位表示版本号
        Unsigned  char Headlen;         //一个字节后 4 个比特位表示首部长度
    }
    unsigned char ip_tos;               //8 位服务类型
    unsigned short ip_totallength;      //16 位总长度(字节)
    unsigned short ip_id;               //16 位标识
    union
    {
        Unsigned short flags;           //两个字节前 3 个比特位表示标志
        Unsigned short offset;          //两个字节后 13 比特位表示片偏移
    }
    unsigned short ip_ttl;              //8 位 TTL
    unsigned short ip_protocol;         //8 位协议类型(TCP、UDP 或其他)
    unsigned short ip_checksum;         //16 位 IP 分组首部简单校验和
    unsigned long ip_srcaddr;           //32 位源 IP 地址
    unsigned long ip_destaddr;          //32 位目的 IP 地址
}IPHeader, *ptrIPHeader;


int main()
{
    //初始化要发送的数据
    const int BUFFER_SIZE = 1000;
    char icmp_buffer[BUFFER_SIZE];
    Unsigned short checkBuffer[65535];
    const char *icmpData = "create ICMP echo request packet and send!";

    //初始化 Windows Socket DLL
    WSADATA wsd;
    if (WSAStartup(MAKEWORD(2, 2), &wsd) != 0)
    {
        printf("WSAStartup() failed: %d ", GetLastError());
```

```
      return -1;
}
//创建Raw Socket
SOCKET rawSocket = WSASocket(AF_INET, SOCK_RAW, IPPROTO_IP, NULL, 0,
                WSA_FLAG_OVERLAPPED);
if (rawSocket == INVALID_SOCKET)
{
   printf("WSASocket() failed: %d ", WSAGetLastError());
   return -1;
}


//设置控制选项
*DWORD bOption = TRUE;
int retResult= setsockopt(rawSocket, IPPROTO_IP, IP_HDRINCL, (char*)
             &bOption, sizeof(//bOption));
if (retResult == SOCKET_ERROR)
{
   printf("setsockopt(IP_HDRINCL) failed: %d ", WSAGetLastError());
   closesocket(rawSocket);
   WSACleanup();
   return -1;
}
//设置接收超时时间
Int recv_timeout = 1000;
int retResult= setsockopt(rawSocket, SOL_SOCKET, SO_RECVTIMEO, (char*)
             & recv_timeout, sizeof(recv_timeout));
if (retResult == SOCKET_ERROR)
{
   printf("setsockopt( SO_RECVTIMEO) failed: %d ", WSAGetLastError());
   closesocket(rawSocket);
   WSACleanup();
   return -1;
}
sockaddr_in  remote_addr;
int remote_addr_len = sizeof(remote_addr);
memset(&remote_assr,0,sizeof(remote_addr));
char *recvbuf = new char[MAX_PACKET + sizeof(tagIPHead)];

int nRecv = recvfrom(rawSocket,recvbuf,MAX_PACKET + sizeof(tagIPHead),0,
         (struct sockaddr *)&remote_addr,&remote_addr_len );
Unsigned short ip_packet_size;
ptrIPHeader = (tagIPHeader *)recvbuf;
ip_packet_size = (ptrIPHeader->HeadLen&0x0f) * 4;
ptrICMPHeader = (tagICMPHeader *)(recvbuf + ip_packet_size);
//ICMP报文首部字段解析结果
```

```
        sprintf("the type of icmp packet:%d\n", ptrICMPHeader -> type);
        sprintf("the code of icmp packet:%d\n", ptrICMPHeader -> code);
        sprintf("the type of checksum packet:%h\n", ptrICMPHeader -> icmpheadcheck);
        sprintf("the idtype of icmp packet:%d\n", ptrICMPHeader -> id);
        sprintf("the sequence of icmp packet:%d\n", ptrICMPHeader -> seq);

        //关闭 Socket，释放资源
        closesocket(rawSocket);
        WSACleanup();
        return   true;
   }
```

第 8 章 传输层 UDP 协议分析与实践

8.1 引　　言

UDP 协议提供一种面向不同终端的不同进程间的无连接传输服务，由于这种服务不确认报文是否到达，不对报文进行排序，也不进行流量控制，因此 UDP 用户数据报在网络中传输时，可能会出现丢失或比特差错等。对于差错控制、流量控制的处理，由上层协议（应用层）根据需要自行解决，UDP 协议本身并不提供可靠传输。与 TCP 协议相同的是，UDP 协议也是通过端口号支持多路复用功能的。

由于 UDP 协议通信开销很小，效率比较高，因此比较适用于对可靠性要求不高，但需要快捷、低延迟通信的应用场合，如多媒体通信等；适用于实时通信应用系统（如 IP 电话、视频会议等），要求主机以恒定速率发送 UDP 用户数据报，并允许一定程度的 UDP 报文丢失，但不允许端到端传输有太大延迟或延迟抖动时间。在航空航天领域，某些应用既需要实时性，又需要可靠性，解决方法是在传输层采用 UDP 协议，在不影响实时通信的前提下，在应用层增加一些简单的可靠通信机制，如对丢失报文进行简单重发；如果直接采用 TCP 协议，虽然可靠性得到了保证，但由于 TCP 协议太复杂，而且系统资源开销比较大，因此无法保证应用实时性。

UDP 协议通信的特点是无连接，即源节点在发送 UDP 用户数据报之前不需要建立连接，数据传输结束后，也不需要释放连接，减少了处理开销和发送延迟。UDP 协议尽最大努力交付，不保证数据可靠性，这样源节点不需要维护许多参数、定时器、复杂连接状态表等资源，简化了 UDP 协议的复杂性。UDP 协议支持一对一、一对多、多对一和多对多的交互通信。UDP 协议不提供拥塞控制，因此当网络中发生拥塞时，发送方不会因此降低发送速率；TCP 协议由于具有拥塞控制机制，因此会适当地降低发送方发送速率，结果会造成网络中存在大量 UDP 用户数据报，产生 UDP/TCP 协议通信的不公平现象。

8.2　UDP 协议工作原理

8.2.1　UDP 协议语法及语义

源节点 UDP 协议对应用进程交付下来的协议数据单元，添加 UDP 首部后就向下交付给 IP 协议。目的节点 UDP 协议对 IP 层交付上来的 UDP 用户数据报首部进行简单处理，去除首部后将数据部分按照目的端口号交付给上层协议的应用进程，一次交付一个完整的报文。交付给上层协议的应用进程产生的协议数据单元大小要合适，如果太大，则在 IP 层需要分片，如果太小，则传输层首部相对太大，降低了网络传输效率。UDP 用户数据报由首部字段和数据字段两部分组成，其中首部字段有 8 个字节，共 4 个字段组成，每个字段都是 2 个字节，如图 8-1 所示。

（1）源端口号：发送方的 UDP 端口号，支持 UDP 多路复用机制。

（2）目的端口号：接收方的 UDP 端口号，支持 UDP 多路复用机制。

（3）UDP 用户数据报长度：UDP 首部+数据，以字节为单位，最小为 8 个字节（UDP 首部长度）。

（4）在计算校验和时，临时把伪首部和 UDP 用户数据报连接在一起计算该字段的值：伪首部+（UDP 首部+数据）；伪首部只是为了计算校验和临时产生的，计算结束后便丢弃。伪协议头包含源 IP 地址、目的 IP 地址、协议号及 UDP 用户数据报长度等字段，有关信息来自 IP 报头；校验和是可选字段，当该字段为 0 时，表示发送方没有为该 UDP 用户数据报提供校验和。

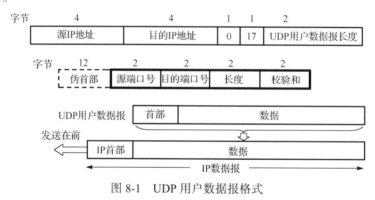

图 8-1 UDP 用户数据报格式

8.2.2 UDP 协议时序关系

UDP 用户数据报是通过 IP 协议进行发送和接收的；网间寻址由 IP 地址来完成，进程间寻址由 UDP 端口来实现。在发送数据时，UDP 实体构造好一个 UDP 用户数据报后递交给 IP 协议，IP 协议将整个 UDP 用户数据报封装在 IP 分组中，即加上 IP 分组首部，形成 IP 分组发送到网络上。在接收数据时，UDP 实体先判断接收到的 UDP 用户数据报的目的端口是否与当前使用的端口匹配，如果匹配，则将 UDP 用户数据报放入相应的接收队列；否则丢弃该 UDP 用户数据报，并向源端口发送一个"端口不可达"的 ICMP 报文。另外，当接收缓存区已满时，即使是端口匹配的 UDP 用户数据报也要丢弃，如图 8-2 所示。

发送 UDP 用户数据报：发送方应用进程发送数据到出队列（有边界），UDP 协议按照队列先进先出（FIFO）顺序，从队列中取出数据（按照应用层交付数据大小），并在加上 UDP 用户数据报首部后，将 UDP 用户数据报交付给 IP 层。IP 层将整个 UDP 用户数据报封装在 IP 分组中再交付给下层。如果出队列溢出，UDP 协议需要告知应用层用户进程暂停发送数据。网间寻址由 IP 地址来完成，进程间寻址则由 UDP 端口来实现。一个用户进程与多个远程进程通信可使用一个出队列。

接收 UDP 用户数据报：UDP 实体在接收 UDP 用户数据报时，首先检查校验和字段（可选）是否出错，如果出错，则丢弃数据但不发送 ICMP 差错报告报文；否则，UDP 实体根据目的端口号查找对应的入队列。如果找到相应的入队列，则将数据放入相应接收队列，并向上层按照原始数据大小交付；否则丢弃，并向源端口发送一个"端口不可达"ICMP 差错报告报文；当接收队列已满时，新接收到的数据

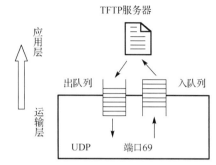

图 8-2　发送和接收 UDP 用户数据报示意图

要丢弃，并向源端口发送一个"端口不可达"ICMP 差错报告报文；当多个用户向一个服务器发送 UDP 用户数据报时，服务器在传输层使用一个 UDP 接收入队列统一接收并管理数据。

8.3　UDP 协议发送 UDP 用户数据报

UDP 协议发送 UDP 用户数据报源代码如下。

```
#include "Winsock2.h"
#include "stdio.h"
#include "Ws2tcpip.h"

#pragma comment(lib,"ws2_32.lib")

//定义UDP伪首部数据结构，计算UDP首部校验和时用
Typedef struct tagPseUDPHeader
{
    Unsigned int srcAddress;        //源IP地址
    Unsigned int dstAddress;        //目的IP地址
    Unsigned char reserved;         //保留
    unsigned char protocolType;     //协议类型
    Unsigned short UDPTotalLen;     //UDP用户数据报总长度（首部+数据）
}PseUDPHeader,* PseUDPHeader;

//定义UDP用户数据报首部数据结构
typedef struct    tagUDPHeader
{
    unsigned short src_port;        //16位源端口号
    unsigned short dst_port;        //16位目的端口号
    unsigned short udp_length;      //16位总长度（首部+数据）
    unsigned short udp_checksum;    //16位简单校验和
}UDPHeader, *ptrUDPHeader;
```

```c
//定义 IP 分组首部数据结构
typedef struct    tagIPHeader
{
    unsigned char ip_verlen;              //4 位首部长度+4 位版本号
    unsigned char ip_tos;                 //8 位服务类型
    unsigned short ip_totallength;        //16 位总长度（字节）
    unsigned short ip_id;                 //16 位标识
    unsigned short ip_offset;             //3 位标志+片偏移
    unsigned short ip_ttl;                //8 位 TTL
    unsigned short ip_protocol;           //8 位协议类型（TCP、UDP 或其他）
    unsigned short ip_checksum;           //16 位 IP 分组首部简单校验和
    unsigned long ip_srcaddr;             //32 位源 IP 地址
    unsigned long ip_destaddr;            //32 位目的 IP 地址
}IPHeader, *ptrIPHeader;

int main()
{
    //初始化要发送的数据
    const int BUFFER_SIZE = 1000;
    char udp_buffer[BUFFER_SIZE];
    Unsigned short checkBuffer[65535];
    Char send_buffer[65535];
    const char *udpData = "create UDP datagram and send!";

    //初始化 Windows Socket DLL
    WSADATA wsd;
    if (WSAStartup(MAKEWORD(2, 2), &wsd) != 0)
    {
        printf("WSAStartup() failed: %d ", GetLastError());
        return -1;
    }
    //创建 Raw Socket
    SOCKET rawSocket = WSASocket(AF_INET, SOCK_RAW, IPPROTO_IP, NULL, 0,
                    WSA_FLAG_OVERLAPPED);
    if (rawSocket == INVALID_SOCKET)
    {
        printf("WSASocket() failed: %d ", WSAGetLastError());
        return -1;
    }

    //设置发送超时时间
    Int timeout = 1500;
    Setsockopt(RawSocket, SOL_SOCKET, SO_SNDTIMEO, (char *)&timeout, sizeof(timeout));

    //设置首部控制选项，使用 IP_HDRINCL
```

```
DWORD bOption = TRUE;
int retResult=setsockopt(s, IPPROTO_IP, IP_HDRINCL, (char*) &bOption, sizeof(bOption));
if (retResult == SOCKET_ERROR)
{
    printf("setsockopt(IP_HDRINCL) failed: %d ", WSAGetLastError());
    closesocket(rawSocket);
    WSACleanup();
    return -1;
}
//填充 IP 分组首部各字段
IPHeader = {0};
IPHeader.ip_version = (0x04<<4|sizeof(IPHeader)/4);
IPHeader.ip_tos = 0;
IPHeader.ip_totallength = sizeif(IPHeader) + sizeof(TCPHeader) + sizeof(tcp_data);
IPHeader.ip_id = 0;
IPHeader.ip_ttl = 64;
IPHeader.ip_protocol=IPPROTO_UDP;
IPHeader.ip_checksum = 0
IPHeader.ip_srcaddr = inet_addr("192.168.0.1");
IPHeader.ip_dstaddre = inet_addr("192.168.0.2");
//初始化 UDP 用户数据报首部各字段
const u_short uRemotePort = 8000;
UDPHeader.dst_port = htons(uRemotePort);          //接收方端口
const u_short uLocalPort = 1000;
UDPHeader.src_port = htons(uLocalPort);           //发送方端口
const unsigned short iUdpSize = sizeof(udpHeader) + strlen(strMessageSend);
UDPHeader.udp_length = htons(iUdpSize);
UDPHeader.udp_checksum = 0;
//计算 UDP 用户数据报首部简单校验和字段
Memset(checkBuffer, 0, 65535);
Memcpy(checkBuffer, &pseUDPHeader,sizeof(pseUDPHeader));
Memcpu(checkBuffer + sizeif(pseUDPHeader), & UDPHeader,sizeof(UDPHeader));
Memcpu(checkBuffer + sizeif(pseUDPHeader) + sizeif(UDPHeader),
&udpData,sizeof(udpData));
UDPHeader .checksum = checksum(checkBuffer, sizeif(pseUDPHeader) +
        sizeif(UDPHeader)
+ sizeof(udpData));

//构建待发送的 UDP 用户数据报
Memset(udp_buffer, 0, 1000);
Memcpy(udp_buffer, &IPHeader, sizeof(IPHeader));
Memcpy(udp_buffer + sizeof(IPHeader),&UDPHeader, sizeof(UDPHeader));
Memcpy(udp_buffer +size0f(IPHeader) +sizeof(UDPHeader),udpData,sizeof(udpData));
//填充 Socketaddr_in 数据结构
sockaddr_in  remote_addr;
remote_addr.sin_family = AF_INET;
```

```
    remote_addr.sin_port = htons(8000);
    remote_addr.sin_addr.s_addr = inet_addr("192.168.0.6");
//发送UDP用户数据报
    int iSendSize = size0f(IPHeader) +sizeof(UDPHeader) + sizeof(udpData);
    int  retResult = sendto(s, buffer, iSendSize, 0, (struct sockaddr *)
                    &remote_addr,
    sizeof(remote_addr));
        if (retResult == SOCKET_ERROR)
            printf("sendto() failed: %d ", WSAGetLastError());
        else
            printf("sent %d bytes ", retResult);

//关闭Socket,释放资源
    closesocket(rawSocket);
    WSACleanup();
    return   true;
}
```

8.4　UDP 协议接收 UDP 用户数据报

UDP 协议接收 UDP 用户数据报源代码如下。

```
#pragma comment(lib, "ws2_32.lib)
#include "stdio.h"
#include "winsock2.lib"
#include "process.h"

Int main(int argc, char *argv[])
{
  WSAData wsaData;
Int iReturn = 0;
//初始化Windows Socket DLL
WSAStartup(WINSOCK_VERSION,&wsaData);
//创建Socket
SOCKET SocketServer = socket(AF_INET,SOCK_DGRAM, 0);
Struct sockaddr_in serverAddr
Memset(&serverAddr, 0, sizeof(serverAddr));
serverAddr.sin_family = AF_INET;
serverAddr.sin_port = htons(8888);
serverAddr.sin_addr.s_addr  = INADDR_ANY;
//绑定UDP服务器端端口号
iReturn = bind(SocketServer,(struct sockaddr *)&serverAddr,sizeof(serverAddr));
if(iReturn == SOCKET_ERROR)
{
    Return -1;
}
```

```
struct sockaddr_in  * clientAddr
int fromlen;
char buffer[1024];
//接收UDP用户数据报
While(true)
{
  clientAddr = (struct sockaddr_in *)HeapAlloc(GetProcessHeap(),HEAP_ZERO_MEMORY,
            sizeof(struct sockaddr_in));
  memset(clientAddr, 0, sizeof(struct sockadd_in));
  fromlen = sizeof(struct sockaddr_in);
  iReturn = recvfrom(SocketServer, buffer, sizeof(buffer),0,
           (struct sockaddr *)clientAddr, &clientAddr);
  If(iReturn == SOCKET_ERROR)
  {
    break
  }
  _beginthread(serverThread,0,clientAddr);
  }
}

Void serverThread(LPVOID lpParam)
{
  Struct sockaddr_in *SocketClientAddr = (struct sockaddr_in *)lpParam;
  SOCKET sockServerthread = socket(AF_INET, SOCK_DGRAM, 0);
  Char *buf = "send to client from server";
  Sendto(sockServerthread, buf, sizeof(buf),0,
       (struct sockaddr *) SocketClientAddr, & SocketClientAddr );
  HeapFree(GetProcessHeap(),0, SocketClientAddr);
  Closesocket(sockServerthread);
}
```

第 9 章 传输层 TCP 协议分析与实践

9.1 TCP 协议概述

TCP/IP 协议于 20 世纪 70 年代末开始研究和开发,以此作为 ARPANET 网络的第二代协议。ARPANET 是美国于 1969 年建立的世界上第一个计算机网络。1983 年初,ARPANET 完成了向 TCP/IP 协议全部转换的工作;同年,美国加利福尼亚大学伯克利学院推出了内含 TCP/IP 协议的第一个 BSD UNIX,大大推动了 TCP/IP 协议的应用和发展。现在,TCP/IP 协议已广泛应用于各网络中,不论是局域网还是广域网,都可以使用 TCP/IP 协议来构造网络环境。除了 UNIX、Windows、NetWare 等一些著名的网络操作系统,其他网络都将 TCP/IP 协议纳入了其体系结构中。以 TCP/IP 协议为核心协议的因特网促进了 TCP/IP 协议的应用和发展,其已成为事实上的国际标准。

从网络体系结构来看,TCP/IP 协议集是 OSI 参考模型 7 层结构的简化,如图 9-1 所示,它将网络体系结构划分为 4 层:应用层、传输层、网络层和网络接口层。其中,网络接口层提供了与物理网络接口相关的方法和规范,对应 OSI 参考模型的物理层和数据链路层;网络层与 OSI 参考模型的网络层对应;传输层与 OSI 参考模型的传输层对应;应用层包含了 OSI 参考模型的会话层、表示层和应用层功能,规定了与特定的网络应用相关的应用协议,如远程登录、文件传送及电子邮件等。

OSI	TCP/IP协议集	
应用层	应用层	Telnet、FTP、SMTP、DNS、HTTP 及其他应用协议
表示层		
会话层		
传输层	传输层	TCP、UDP
网络层	网络层	IP、ARP、RARP、ICMP
数据链路层	网络接口层	各种通信网络接口(以太网等)(物理网络)
物理层		

图 9-1 TCP/IP 协议集

传输层为网络上不同终端应用进程之间提供端到端的逻辑通信;网络层为主机之间提

供逻辑通信；传输层向高层用户屏蔽下层网络核心的细节，如网络拓扑、采用的路由选择协议等，使应用进程看见的就好像是在两个传输层实体之间有一条端到端的逻辑通信信道。传输层 TCP 协议对收到的报文段进行差错检测。传输层需要两种不同的协议，即面向连接的 TCP 协议和无连接的 UDP 协议。

虽然 TCP 协议是 TCP/IP 协议集中的一个网络通信协议，但它有很大的独立性，对下层网络协议只有基本的要求，很容易在不同的网络上应用，因此可以在众多的网络上工作。TCP 协议提供可靠的面向数据流的传输服务，用户数据可以有序而可靠的进行传输。在分组发生丢失、破坏、重复、延迟或失序的情况下，TCP 协议通过一种可靠的进程间通信机制能够自动纠正各种差错。TCP 协议可以支持许多高层协议（Upper Level Protocol，ULP），它对高层协议的数据结构无任何要求，只将它们作为一种连续的字节流处理。TCP 协议对分组没有太多的限制，最大分组为 65KB，较大的分组将在 IP 层分成分片进行传送，但一般 TCP 的实现都规定了数据段的长度。此外，TCP 协议在数据流中加入了一个面向字节的序号，以便管理发送方和接收方 TCP 实体间连续的数据流。

TCP 协议的主要功能是在一对进程之间提供传输层面向连接的通信服务，如图 9-2 所示。TCP 连接管理可以分为 3 个阶段：建立连接、数据传输和终止连接。TCP 协议主要通过套接字为上层协议提供面向连接的 API 传输服务，利用套接字可使一个上层协议主动发起与另一个上层协议之间的唯一连接。当连接一旦建立起来，并且该连接处于活动状态时，TCP 协议可以产生并发送分组。当发送结束后，连接双方都要终止各自的连接。为了保证提供可靠的服务，TCP 协议还提供了确认、流量控制、拥塞控制、端口复用及同步等功能。TCP 协议是面向连接的传输层协议，在通信前通过 3 次握手建立连接，在通信过程中对该连接进行维护，在通信结束后释放该连接。TCP 连接是一个虚的逻辑连接，不是真正的物理连接，每一条 TCP 连接只能有两个端节点（end-point）；每一条 TCP 连接只能是一对一通信（单播通信）。TCP 协议通过以下 3 个机制提供传输层可靠通信服务。

（1）差错控制：序号 + 确认反馈 + 超时重传。
（2）流量控制：采用可变大小的滑动窗口机制。
（3）拥塞控制：慢启动、拥塞避免、快速重传、快速恢复。

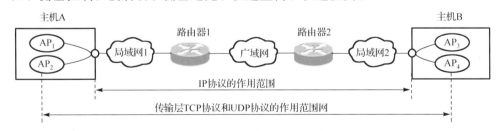

图 9-2　TCP 协议应用范围

TCP 协议提供全双工面向字节流通信，对 TCP 报文段中数据部分的每一个字节都编有序号。TCP 协议对应用进程一次把多大数据块交付到 TCP 协议发送缓存中并不关心，假设发送方应用进程可能交付给传输层 TCP 协议 10 个数据块，但 TCP 协议有可能只用了 3 个 TCP 报文段就发送完了，如图 9-3 所示。TCP 会根据接收方给出的窗口值、当前网络拥塞程度和最大报文段长度（MSS）来决定一个报文段应包含多少个字节，但 UDP 协议发送的报文段长度是应用进程已确定的。TCP 协议可以把发送缓存中的一个大数据块分割为若干个 TCP 报文段后再发送，也可以等待发送缓存中积累足够多的数据后，再构成一个 TCP 报文段发送出去。

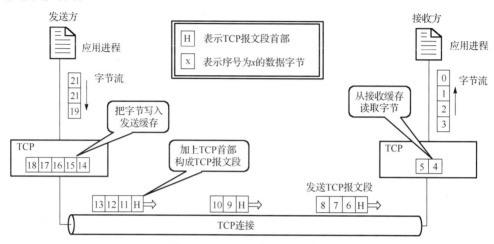

图 9-3　TCP 基于字节流的传输

当一个应用进程需要与一个远程进程进行通信时，因为主机上存在多个进程，所以它必须指明要与哪个进程进行通信。因此，需要有一种机制对不同进程进行编号，而 TCP 协议采用端口号。在网络环境下标识一个应用进程可以采用二元组：IP 地址 + 端口号。传输层对不同应用进程的编号（编址）为端口号，术语称为传输层服务访问点（Transport Service Access Point，TSAP）。在此情况下，多个（本地或异地）进程可以同时与一个远程进程进行通信，反过来，一个本地进程也可以同时与多个远程进程进行通信。发送方应用层的各种应用进程都能将其数据通过不同源端口号向下交付给传输层的一个通信实体，称为发送方端口复用；接收方传输层根据接收到的数据单元首部中的不同目的端口号向上交付给应用层相应的应用进程，称为接收方端口分用，如图 9-4 所示。TCP 的多路复用功能是通过端口复用和端口分用机制来提供的。不同主机上的 TCP 实体对端口地址的选取是独立的，只要保证本地端口地址的唯一性，就可以保证整个 TCP 连接的唯一性，即网络间唯一的 IP 地址和本地唯一的端口地址将确定唯一性通信连接的一端。

TCP 协议提供面向连接服务，虽然网络层提供尽最大努力交付的不可靠通信服务，但 TCP 实体通过软件在传输层提供一条全双工的端到端可靠通信信道，如图 9-5 所示。传输层两个对等实体在通信时传送的数据单位叫作传输层协议数据单元（Transport Protocol Data Unit，TPDU）。两端节点必须调用同一个传输层协议实体；TCP 协议产生的协议数据单元是 TCP 报文段，UDP 协议产生的协议数据单元是 UDP 用户数据报。TCP 协议要提供可靠的传输服务，不可避免地增加了许多开销：TCP 报文段首部比较大，在通信时需要占用大量的主机资源。TCP 协议不提供广播或组播服务，仅支持单播通信。UDP 协议提供无连接服务，通信双方发送 UDP 用户数据报之前不需要先建立连接；接收方接收到 UDP 用户数据报后，只进行检错，不给出任何确认。虽然 UDP 协议不提供可靠通信，但其满足了一定的实时通信服务需求，主要用于多媒体通信，支持单播、广播和组播通信。

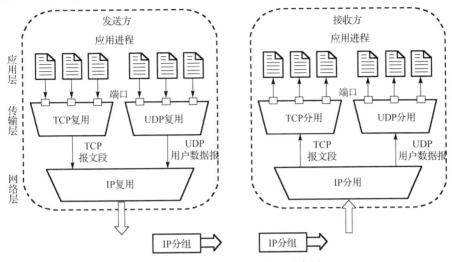

图 9-4　端口复用和端口分用示意图

TCP 协议在通信时首先要通过 3 次握手建立一个 TCP 连接，建立连接是在传输层完成的，路由器无法感知；TCP 报文段和 UDP 用户数据报均以 IP 分组的形式在网络上传输，每一个 IP 分组在网络中是独立路由，所以 IP 分组存在丢失、重复、延迟、差错和乱序等传输问题。IP 分组的可靠性只能由传输层 TCP 协议负责。

在因特网上，计算机使用的操作系统的种类很多，不同的操作系统又使用不同的规范对进程进行标识编码。在网络环境下，为了使运行不同操作系统的计算机应用层进程能够相互通信，就必须使用统一的规范对应用层进程进行标识编码。传输层对应用层进程进行统一编号（编址）：端口号是一个 16 比特位二进制数，其具有本地含义，即端口号在本地是唯一的；同一个主机，针对同一传输层 TCP 协议，不能有两个或两个以上相同的端口号；不同主机可能具有相同端口号，但代表不同主机的不同应用层进程，相互没有影响。

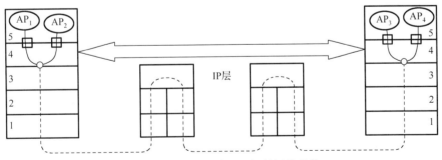

图 9-5　不同端系统进程间的网络通信

端口号是用来标识不同应用层进程的，但在进行网络通信时，需要同时提供目的 IP 地址+目的端口号和源 IP 地址+源端口号+协议类型（TCP 协议或 UDP 协议）。因为网络上不同进程之间一般采用全双工通信模式，除发送方发送数据给接收方之外，接收方也需要发送数据（或应答）给发送方；所以在网络中一般采用五元组标识通信双方：源 IP 地址、源端口号、目的 IP 地址、目的端口号、协议类型。

端口号一般分为服务器端端口号和客户端端口号。其中，服务器端端口号包括周知端口号和注册端口号。周知端口号的数值一般为 0~1023，由 IANA 组织管理和分配；注册端口号的数值一般为 1024~49151，为没有周知端口号的应用服务器程序使用，这个范围的端口号必须在 IANA 登记，以防重复使用。客户端端口号（或动态端口号）：数值范围为 49152~65535，留给客户端应用程序暂时使用；由操作系统临时分配，在接收通信后由操作系统收回。当服务器进程收到客户进程的报文时，就知道了客户进程使用的动态端口号；在通信结束后，客户端端口号立即释放，可供其他客户进程使用。

每个主机上的应用进程通过不同端口号标识，使用单一的传输层实体进行通信，每个应用进程只需保证自己的端口号的本地唯一性就可以保证整个通信连接的唯一性。服务器端和客户端使用端口号的方法不同，服务器端一般使用周知端口号等候客户端的连接或其他服务；客户端只需操作系统分配其一个临时动态端口号即可。在一次数据通信结束后，客户端应用程序释放占用的动态端口号，该动态端口号可以分配给其他客户端应用程序使用；服务器端一般是一直工作，使用的周知端口号保持不变。

9.2 TCP 协议工作原理

TCP 协议通信的基本传输单元为段或 TCP 报文段，一个 TCP 报文段由首部和数据两部分组成。TCP 数据部分是无结构的字节流，字节流中的数据是由一个个字节序列构成的，无任何可供解释的结构，这一特征使得 TCP 报文段的长度是可变的。因此，TCP 协议中的序号和确认序号都是针对字节流中字节的，而不是针对 TCP 报文段的。为了保证传输的可靠性，接收方 TCP 实体要对发送方 TCP 实体传来的 TCP 报文段中的数据字节给予确认。在一般情况下，接收方将确认已正确收到的连续字节流的前 n 个字节，给出的确认序号指示的是下一个（$n+1$）希望接收的字节。这种面向字节的累计确认方式的优点是，在可变长度的数据段传输方式下，不会发生确认的二义性，避免了因部分应答丢失而产生的问题，在一定程度上减少了网络传输开销，并且实现起来也比较容易。两个使用 TCP 协议进行通信的对等实体间的一次通信过程，一般要经历建立连接、数据传输（双向）和终止连接等阶段。TCP 协议内部通过一套完整的状态转换机制来保证各阶段的正确执行，为上层应用提供双向、可靠、顺序及无重复的数据流传输服务。

9.2.1 TCP 协议语法及语义

TCP 协议产生的 TCP 报文段由首部和数据两部分组成，其中首部由 20 字节固定首部和 0~40 字节可变首部组成，所以 TCP 报文段首部字节的变化范围为 20~60 字节，如图 9-6 所示。

（1）源端口号、目的端口号：各占 2 字节。端口号是传输层提供给应用层的 TSAP 地址；传输层的端口复用和端口分用功能都要通过端口号来实现。

（2）序号：占 4 字节，TCP 连接中传送的数据流为字节流，字节流中的每一个字节都编有一个序号。序号标识 TCP 报文段数据部分的第一个字节在字节流中的序号,如图 9-7 所示。如果通信双方协商发送的初始序号为 0,则需要发送的数据最大长度限定为 100 字节。如果字节流中的第 1 个字节序号从 0+1=1 开始,则 4 个 TCP 报文段序号分别为 1、101、201 和 301。如果通信双方协商发送的初始序号为 2000,则需要发送的数据最大长度限定为 100 字节;如果字节流的第 1 个字节序号从 2000+1=2001 开始,则此时 4 个 TCP 报文段序号分别为 2000+1、2000+101、2000+201 和 2000+301。

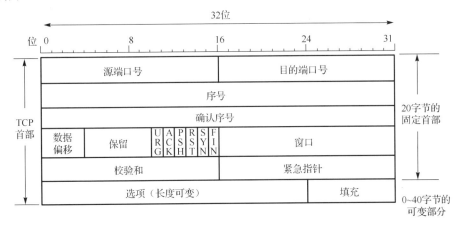

图 9-6 TCP 报文段首部结构

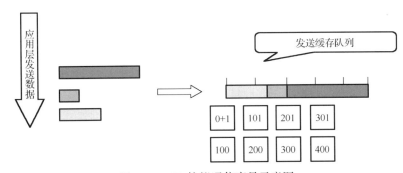

图 9-7 TCP 协议通信序号示意图

在每条 TCP 连接上传送的每个数据都有一个与之相对的序号,这是 TCP 实体的重要概念之一。以字节为单位递增的 TCP 序号主要用于数据排序、重复检测、差错处理及流量控制窗口等 TCP 协议机制上,保证了传输任何数据的可靠性。由于 TCP 序号字段占 4 个字节,表示的序号空间范围为 $0 \sim 2^{32}-1$,因此发送字节的序号编码算法都要以 2^{32} 为模。TCP 序号不仅用于保证数据传送的可靠性,还用于保证建立连接(SYN 段请求)和终止连接(FIN 请求)的可靠性,每个 SYN 报文和 FIN 报文都要消耗一个序号。

（3）确认序号：占 4 字节，表示接收方希望收到发送方发送的下一个 TCP 报文段数据部分的第一个字节序号，并且在 ACK = 1 的情况下才有效。如果确认序号为 N，则表示 $N-1$ 号以前的数据字节接收方已经成功接收到，期望接收的下一个 TCP 报文段数据部分的第一个字节序号为 N，实际也标识期望接收的下一个 TCP 报文段序号为 N。例如，主机 B 收到主机 A 发送的一个 TCP 报文段，其序号为 501，数据部分长度为 200 个字节，标识该数据部分序号范围为 501～700，则此时主机 B 期望收到主机 A 的下一个 TCP 报文段序号应该为 701，而不是 501 或 700，所以确认序号应该为 701。

（4）数据偏移（首部长度）：占 4 比特位，表示 TCP 报文段首部长度，以 4 字节为计算单位，最大值为 60 字节。TCP 报文段首部长度范围为 20～60 个字节，其中选项字段范围为 0～40 字节。

（5）保留：占 6 比特位，保留字段是为以后使用的，目前全为 0。

（6）比特标志位：UAPRSF 6 个标志字段占 6 个比特位。

① 紧急比特（URG）：当 URG = 1 时，表明紧急指针字段有效，TCP 报文段中有紧急数据，应尽快传送（相当于高优先级的数据）。紧急数据在数据字段的位置由紧急指针（urgent pointer）字段给出。

② 确认比特（ACK）：当 ACK = 1 时，确认序号字段有效；当 ACK = 0 时，确认序号字段无效。

③ 推送比特（PSH）：PUSH 操作，如果发送方设置 PUSH=1，则表明要进行 PUSH 操作，此时发送缓存区即使有发送窗口限制，数据也要立即发送；接收方 TCP 协议在收到 PUSH=1 的数据时，不需要在接收缓存区中排队，可尽快交付给应用进程。TCP 协议中发送方接收到上层数据后，先缓存到出队列，然后根据发送窗口大小分段发送数据。若应用层在数据发送完成前出现异常情况，如突然关机，则发送方无法知道自身递交的数据是否已被 TCP 协议发送。

上述情况的解决方法是发送方 TCP 实体接收到上层 PUSH 操作后，TCP 会将出队列中的所有数据迅速地分段发送出去，而不受发送方当前发送窗口大小的限制。接收方 TCP 实体接收到带 PSH 标志的 TCP 报文段后，将迅速把这些数据段递交给上层协议，而不用在接收缓存中排队。

由此可见，在一个 TCP 连接上进行正常数据传输的过程中，发送方会按规定的方式来分段和发送上层协议递交的数据。对于递交了该数据的上层协议来说，在数据发送完成前，它是无法知道自身递交的数据是否在发送，或者发送是否由于本地或远地 TCP 实体正等待一个合适的窗口而被延迟。推进数据机制可使上层协议递交的数据迅速地从本地推向远地，而不受发送方当前发送窗口大小和发送方式的限制。

④ 复位比特（RST）：当 RST = 1 时，表明 TCP 连接中出现严重差错，应对差错连接进行复位，重新回到初始状态。

⑤ 同步比特（SYN）：当 SYN = 1 时，表明这是一个 TCP 连接请求。

⑥ 终止比特（FIN）：当 FIN = 1 时，表明发送方的数据已发送完毕，要求释放 TCP 连接，该 TCP 报文段是一个释放连接请求。

（7）窗口：占 2 字节，表示接收方通知发送方接收缓存的有效大小，单位为字节；主要目的是发送方接收到对方"窗口信息"后，通过调整发送方发送窗口大小来控制发送速率，以实现端到端流量控制。

（8）校验和：占 2 字节。校验和范围包括 TCP 伪首部（临时产生，共 12 个字节）+ TCP 首部 + 数据。TCP 伪首部包括源 IP 地址+目的 IP 地址+1 字节保留（全 0）+协议号（TCP 协议号=6）+TCP 首部长度字段，共 12 个字节。简单校验和计算方法与 IP 协议（或 UDP 协议）计算校验和的方法相同。

（9）紧急指针：占 16 比特位，表示 TCP 报文段中紧急数据字节数，当 URG=1 时，紧急指针字段有效。紧急数据一般放在 TCP 报文段数据最前面。实际上，序号+紧急指针字段的值–1 得到的值，等于本 TCP 报文段中的紧急数据最后一个字节的序号。

紧急指针是指任何一个 TCP 报文段都可以携带紧急数据，以支持上层协议紧急数据的快速传递与处理。紧急数据有关 TCP 报文段相关字段：URG 标志位和紧急指针。

假设发送方正在发送数据，然后等待接收方处理结果返回，发送方突然发现有错误，希望将此过程异常终止。如果此时发送终止命令（Control+C），那这两个字符必然排在接收方队列队尾，等所有数据处理完才能将这两个字符交付上层。

上述情况的解决方法是发送方将 URG 置 1，告知 TCP 协议该数据为紧急数据，则发送方 TCP 协议将含有紧急数据的数据块安排在发送队列队首，以便尽快发送出去；接收方 TCP 协议收到 URG=1 的报文段后，利用紧急指针的值将该报文段的紧急数据提取出来，并不按缓冲排队顺序，而是提前交给应用进程进行处理，以避免对错误数据进行处理。

任何一个输出 TCP 报文段都可以携带紧急数据，以支持上层协议间紧急信息的快速传递。紧急数据必须位于数据段中所有数据的最前端，并在 TCP 报文段头中设置如下字段。

① 设置 URG 标志位，表示当前数据字段中有紧急数据。

② 设置紧急指针。紧急指针与 TCP 报文段的序号之和指向数据字段中紧急数据最后一个字节的位置。对于一个包含该字节的数据字段来讲，其紧急数据长度从段序号开始一直延续到该字节为止。

当接收方 TCP 协议收到含有紧急数据的 TCP 报文段后，首先把紧急数据从正常数据流中分离出来，并保存在适当的地方；然后以一定的方式通知上层协议，希望它尽快地响应和处理；当发送方 TCP 收到上层协议递交的紧急数据后，并不是立即发送紧急数据，而是把它们排入正常数据的发送队列中。此后，发送方 TCP 协议每发送一个数据字段，都要带有 URG 标志和紧急指针，直到出现真正含有紧急数据的数据字段为止，不过越靠后的数据字段的紧急指针的值就越小。也就是说，含有 URG 标志和紧急指针的数据字段不一定就含有紧急数据，只有包含紧急指针指示字节的数据字段才含有真正的紧急数据，即从段序号开始到该字节为止皆为紧急数据。此外，即使发送方 TCP 协议发送窗口为 0，也要对带有 URG 标志的数据字段实行强制发送，不管这些数据字段是否含有紧急数据。

（10）选项：长度可变，为 0~40 字节。TCP 协议最初只规定了一种选项，即最大报文段长度（MSS）；MSS 告诉对方 TCP 实体："我的缓存能接收的 TCP 报文段的数据字段的最大长度是 MSS 个字节。"

① MSS 是 TCP 报文段中的数据字段的最大长度，数据字段加上 TCP 首部才等于整个 TCP 报文段。如果 MSS 太小，则数据传输效率低。假设传送一个字节数据，在网络层数

据传输效率为 1/41（不包括链路层开销）。如果 MSS 太大，则网络处理时间长，开销大，并且在网络层一个大的 TCP 报文段又要分片，构成若干个小 IP 分组，当接收方对小的分组进行重新组装时，如果有一个小的 IP 分片出现差错、丢失或超时，则整个 TCP 报文段都需要重传，所以，MSS 的选择原则一般为 MSS 尽可能大，只要在 IP 层不分片就可以了。

TCP 协议在建立连接时，双方都要将自己可支持的 MSS 写入选项字段进行协商，然后选择最小的 MSS 作为本次通信的最大报文段长度；在进行通信时，两个方向的 MSS 可以不同，如果主机没填写选项 MSS 字段，则 MSS 一般采用默认值 536 字节，传输层默认 TCP 报文段长度为 536+20（固定首部）=556 字节。由于 IP 分组是独立路由，因此不同分组每次选择的路径可能不同，如在某一条路径上按协商好的 MSS 不需要分片，但改选另一条路径可能需要分片，所以最佳的 MSS 很难确定。

② 窗口扩大选项：占 3 字节，其中有一个字节表示移位值 S；新的窗口值等于 TCP 首部中的窗口位数增大到 16+S 比特，S 最大为 14 比特，相当于把窗口值向左移动 S 位后获得的窗口大小。

③ 时间戳选项：占 10 字节，最主要的字段为时间戳字段（4 字节）和时间戳回送应答字段（4 字节）。

发送方在发送 TCP 报文段时，把发送时间写入时间戳字段；接收方在发送确认应答报文段时，将时间戳字段复制到时间戳回送应答字段，发送方可依次计算 RTT。为了解决当前序号绕回到前一个 TCP 连接对应的序号空间问题，如果发送速率为 1GB/s，则不到 35s 序号就会重复。为了使接收方能够将新报文段与延迟报文段分开，可利用时间戳区分。

（11）填充：目的是使整个 TCP 首部长度是 4 字节的整数倍，并按照 32 比特位对齐。

9.2.2 TCP 协议通信的时序关系

TCP 协议通信的 3 个阶段：建立 TCP 连接；数据通信维护连接，用来确保 TCP 报文段传输的可靠性；释放 TCP 连接，这是两个端系统 TCP 实体之间的行为。在建立连接的过程中要解决以下 3 类问题：双方能够确切地知道对方的存在；允许双方协商一些参数（如 MSS、最大窗口大小、服务质量等）；对网络资源，如缓存大小、不同定时器（如重传定时器）、保活定时器和坚持定时器，以及连接状态表等，进行分配并初始化。

TCP 协议通信采用客户/服务器方式。主动发起建立 TCP 连接的应用进程叫作客户进程，简称客户端；被动等待对方建立连接的进程叫作服务器进程，简称服务器端。

1）建立 TCP 连接

在 TCP 协议中，要通过 3 次握手来建立连接，如图 9-8 所示。下面是较常见的 3 次握手建立连接过程。

（1）TCP 实体 A 向 TCP 实体 B 发送一个 TCP 连接请求。例如，该 TCP 报文段简要表示成 $\langle SEQ=x \rangle \langle CTL=SYN \rangle$，同时在选项字段填写 TCP 实体 A 支持的 MSS_1。

（2）TCP 实体 B 将确认 TCP 实体 A 的请求，同时向 TCP 实体 A 发出反向连接请求，一个报文两个作用。例如，该 TCP 报文段为 $\langle SEQ=y \rangle \langle ACKn=x+1 \rangle \langle CTL=SYN, ACK \rangle$，同时填写 TCP 实体 B 支持的 MSS_2。

（3）TCP 实体 A 将确认 TCP 实体 B 的请求，即向 TCP 实体 B 发送确认 TCP 报文段。例如，该 TCP 报文段为〈SEQ=x+1〉〈ACKn=y+1〉〈CTL=ACK〉，有些通信场景可以捎带发送应用数据〈DATA〉，如 HTTP 协议。

（4）TCP 实体 A 在已建立的连接上传输 TCP 报文段。例如，该 TCP 报文段为〈SEQ=x+1〉〈ACKn=y+1〉〈CTL=ACK〉，发送应用数据〈DATA〉。

此外，在建立连接的过程中，对于出现的异常情况，如本地请求与过去遗留在网络中的连接请求序号重复、因系统异常使通信双方处于非同步状态等，TCP 协议要通过使用复位（RST）TCP 报文段来加以恢复，即发现异常情况的一方发送复位 TCP 报文段通知对方来处理异常。

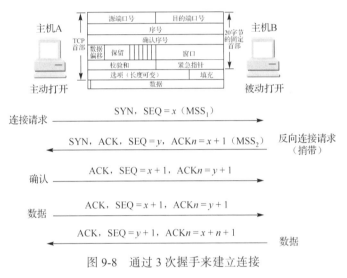

图 9-8 通过 3 次握手来建立连接

TCP 标准规定 1：SYN 置 1 的 TCP 报文段消耗一个序号。

TCP 标准规定 2：单独的确认 TCP 报文段不消耗序号。

2）释放 TCP 连接

由于 TCP 连接采用全双工的数据传输模式，因此一个连接的关闭必须由通信双方共同完成。当通信的一方没有数据需要发送给对方时，可以发送 FIN 报文段向对方申请关闭连接。这时，TCP 连接虽然不再发送数据，但并不排斥在这个连接上继续接收数据。只有当通信的对方也递交了关闭连接请求后，这个 TCP 连接才会完全关闭。实际上，TCP 连接的关闭过程是一个 4 次挥手的过程。

在图 9-9 中，从主机 A 到主机 B 的连接释放了，连接处于半关闭状态。相当于主机 A 向主机 B 说："我已经没有数据要发送了，所以并不能发送数据，如果主机 B 还需要发送数据，那么我仍接收，并给主机 B 发送确认。"

图 9-9 TCP 释放半连接过程

在图 9-10 中，从主机 A 到主机 B 和主机 B 到主机 A，双方的连接均释放了，此时，主机 A 无法向主机 B 发送报文段，主机 B 也无法向主机 A 发送报文段，至此，整个连接已经全部释放。

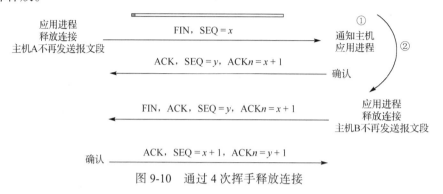

图 9-10　通过 4 次挥手释放连接

命令：netstat –a，可查看本机连接状态。

3）TCP 可靠传输的工作原理

TCP 协议是一种面向连接的传输层协议，提供端到端通信的可靠性，发送方应用进程将数据交付给 TCP 协议后，TCP 协议可无差错的将数据交付给目的应用进程。理想的数据传输需要满足两个条件：一是传输信道不产生差错（主要包括比特差错和传输差错）；二是不管发送方发送速率多快，接收方总来得及将接收的数据交付给上层协议，避免接收缓存溢出。由于以上假设在实际通信环境下很难满足，因此端到端的可靠通信需要解决差错控制、流量控制和拥塞控制的问题，具体方法如下。

（1）差错控制：序号+确认反馈+超时重传。

（2）流量控制：实现发送方发送速率与接收方应用进程从接收缓存区交付速率之间的匹配。一般采用接收方交付速率控制发送方发送速率的方式，如带差错控制的停止-等待协议、连续 ARQ-后退 N 帧协议和连续 ARQ-选择重发协议等。

（3）拥塞控制：主要解决路由器缓冲溢出问题，网络层通过发送源抑制 ICMP 报文来解决；传输层通过端到端拥塞控制技术来解决。

TCP 协议提供面向字节流的传输，将所要发送的数据看成字节流，每一个字节对应一个序号。接收方给发送方发送应答确认也是按照字节而不是 TCP 报文段进行确认，ACK=1，确认序号有效，$ACKn = N$ 表示接收方已接收到 $N–1$ 序号以前的数据。在建立连接时，TCP 协议通信双方首先要协商初始序号。假设发送方和接收方协商的初始序号分别为 x，y，则发送方和接收方开始发送的数据序号分别为 $x+1$，$y+1$。TCP 协议发送的 TCP 报文段首部中的序号字段数值表示该 TCP 报文段的数据字段的第一个字节在数据流（字节流）中的序号。假设主机 A 要发送 5000 字节的文件给主机 B，在建立连接时双方协商 MSS=1000 字节，则在传输层该文件数据被分为 5 个 TCP 报文段进行传输。假设在建立 TCP 连接时，主机 A 随机选择的初始序号 $x=10000$，则数据传输开始序号为 $x+1=10001$。5 个 TCP 报文段首部序号字段的值如下。

（1）第 1 个 TCP 报文段：序号字段=10001（数据部分序号范围为 10001～11000）。

（2）第 2 个 TCP 报文段：序号字段=11001（数据部分序号范围为 11001～12000）。

（3）第 3 个 TCP 报文段：序号字段=12001（数据部分序号范围为 12001～13000）。
（4）第 4 个 TCP 报文段：序号字段=13001（数据部分序号范围为 13001～14000）。
（5）第 5 个 TCP 报文段：序号字段=14001（数据部分序号范围为 14001～15000）。

在 TCP 协议中，ACK 应答采用按字节进行确认的方式，当 ACK=1 时，确认序号字段有效，确认序号等于已收到的数据的最高字节序号+1，或者等于接收方期望下次收到的 TCP 报文段中序号字段的值，或者等于接收方期望下次收到的 TCP 报文段中的数据字段的第一个字节序号。

上述 3 种说法实际上是等价的。TCP 确认应答可采用 3 种方式，一是单独确认：一方给另一方发送一个确认应答，应答中无数据。二是捎带确认：一方在给另一方发送数据时，捎带确认应答信息。三是累积确认：接收方不必对接收的每一个 TCP 报文段逐个发送确认应答，可以在收到几个 TCP 报文段后，对按序到达的最后一个 TCP 报文段发送确认应答即可。累积确认的好处是，某个 ACK 应答丢失不一定会导致发送方重发，所以不需要对该应答重发，也不会影响可靠传输。

假设发送方发送 TCP 报文段的数据部分字节序号的范围为 1201～1300，接收方发送 ACKn=1301，表明 1300 以前的字节均成功接收；发送方再发送下一个 TCP 报文段的数据部分字节序号的范围为 1301～1400，接收方发送 ACKn=1401，表明 1400 以前的字节均成功接收。如果 ACKn=1301 确认丢失，但发送方后来接收到 ACKn=1401 应答，则表明 1400 以前的字节均成功接收，但条件是字节序号为 1301 的 TCP 报文段重发定时器未超时。

TCP 超时重传是指发送方 TCP 实体每发送一个 TCP 报文段，便会启动一个重发定时器，如果重发定时器超时还没有收到对方相应的 ACK 应答，则重传该 TCP 报文段。但是如何确定重发定时器的超时时间是一个问题。如果超时时间设置过大，则发送方 TCP 协议可能在某时刻有许多重传定时器同时工作，会占用大量的系统资源；如果超时时间设置过小，则网络中存在过多的重发 TCP 报文段，会消耗网络可用带宽。所以，超时时间的设置原则是稍大于 RTT。由于网络负载随时变化、分组独立路由，因此很难精确测量到 RTT，目前的解决思路是利用以前的历史 RTT 估算当前的 RTT，具体方法有 Karn/Partridge 算法和 Jacobson/Karels 算法。

4）TCP 可靠通信实现技术

在发送方发送的每一个 TCP 报文段中，首部通过序号字段来说明数据部分的第一个字节在原始字节流中的位置，并启动一个重发定时器。接收方每收到一个 TCP 报文段，首先进行差错检测，如果校验和字段检测正确，则给发送方发送相应的 ACK 应答，TCP 报文段按字节进行确认，否则直接丢弃；如果发送方重发定时器在未超时时收到 ACK 应答，则发送窗口向前滑动，发送方可继续发送后面的 TCP 报文段；如果发送方重发定时器超时，但未收到接收方对该 TCP 报文段的 ACK 应答，则重发 TCP 报文段，此时采用后退 N 帧协议或选择重发机制实施流量控制。

在 TCP 协议差错控制机制中，字节序号占 32 比特位，在 TCP 协议中，ACK 应答都是基于字节而不是基于报文段进行确认的；重发定时器超时时间一般设置为稍大于 RTT；由于网络负载变化、分组独立路由等原因，TCP 连接的 RTT 并不是固定不变的，而是需要使用特定的算法估算较为合理的重传超时时间的。TCP 协议端到端流量控制采用连续 ARQ

协议；TCP 连接的每一端都设有两个缓存：发送缓存和接收缓存。TCP 连接的每一端都各设两个窗口：一个发送窗口和一个接收窗口。TCP 连接两端的两个发送窗口经常处于动态变化之中，一般接收窗口的初始大小相对固定。TCP 报文段首部的窗口字段数值就是接收方接收窗口的有效大小，目的是控制发送方发送窗口数值上限。在理想情况下，发送窗口大小需要小于或等于接收窗口大小；连接一旦建立起来，通信双方就可以在该连接上传输数据了。在数据传输过程中，TCP 协议提供一种基于动态窗口协议的流量控制机制，接收方 TCP 实体能够根据自己当前的接收缓存区的有效容量来控制发送方 TCP 实体传送的数据量。

在建立连接时，双方使用 SYN 报文段或 ACK 报文段中的窗口字段捎带相互通告各自的窗口尺寸。在数据传输过程中，发送方按接收方通告的窗口尺寸和序号发送一定的数据量，接收方可根据接收缓存区的使用状况动态地调整接收窗口的有效大小，并在输出数据段或确认段时捎带将新的窗口尺寸和起始序号（在确认序号字段中指出）通告给发送方。发送方将用新的起始序号和新的接收窗口尺寸来调整发送窗口，接收方也用新的起始序号和新的接收窗口大小来验证每一个输入数据段的可接收性。

问题 1：在 TCP 流量控制中存在死锁现象吗？为什么？如何解决？

在图 9-11 中，发送方 A 根据发送窗口大小，分两次将字节序号 34～43 与 44～53 按照两个 TCP 报文段发送到接收方 B；在正常情况下，接收方给发送方发送两次 ACK 应答，目的是控制发送方的发送窗口向前滑动：ACK=1，ACKn=44，Win=20；ACK=1，ACKn=54，Win=20。此时，接收窗口后沿均应滑动到序号范围 54～63 的位置，发送窗口也应滑动到序号范围 54～63 的位置，这样就可以正常通信了。如果以上两个 ACK 应答在通信过程中全部丢失，就会发生死锁，这时接收方无法向发送方发送 ACK 应答，而发送方等待收到接收方非零窗口通知，接收方等待发送方发送数据。在本例当中，发送重发定时器超时会引起 TCP 报文段重发，由于重发的 TCP 报文段序号小于 54，因此该序号无法落在接收窗口内，所以接收方会将 TCP 报文段丢弃，同时接收方不会给发送方发送 ACK 应答，导致收发双方进入到通信死锁状态。

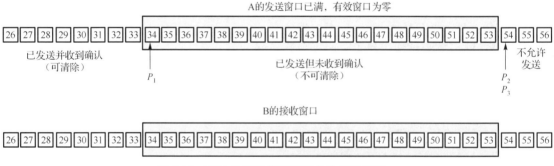

图 9-11　TCP 协议通信死锁现象分析示意图

为了解决死锁问题，发送方引入了坚持定时器。发送方每建立一个 TCP 连接，就启用一个坚持定时器，当发送方检测到发生死锁时，就会启动坚持定时器。死锁的检测条件：发送方收到 ACK 应答，发送窗口无法向前滑动；坚持定时器的初始超时时间设置为重发定时器的两倍；坚持定时器一旦超时，发送方发送一个只有 1 个字节的探测报文，接收方每收到一个探测报文，就必须发送一个 ACK 应答（报告接收方接收窗口的有效大小：确认序号、窗口值）；如果发送方没有收到 ACK 应答，则超时时间加倍，并重发探测报文，一直持续该过程，直到超时时间达到门限值（60s）为止；以后发送方每 60s 发送一个探测报文，直到发送方发送窗口重新打开。

问题 2：你认为 TCP 协议在通信时还存在什么问题？为什么？如何解决？

假设客户端与服务器端建立了一个 TCP 连接，并发送了一些 TCP 报文段，然后客户端出现了故障，使服务器端的 TCP 连接一直处于打开状态，占用了系统资源。当服务器端存在大量半连接状态时，会引起服务器端性能下降或崩溃。

解决上述问题的方法是，在服务器端的每个连接上引入保活定时器。保活定时器的初始超时时间一般设置为 2 小时（7200s），当服务器端收到来自客户端的 TCP 报文段时，该服务器端连接上相应的保活定时器复位；如果服务器端的保活定时器超时，则服务器端不断（每隔 75s）发送一个探测报文给客户端；当服务器端连续发送了 10 个探测报文却还没在该连接上收到客户端应答时，服务器端就关闭该连接。

问题 3：发送方 TCP 协议采用何种机制从发送缓存依据发送窗口取出数据封装并发送？

考虑到 TCP 协议的传输效率，可采用不同的方式来控制 TCP 报文段的发送时机，具体有 3 种方式：第一种是采用触发方式，发送方维持一个变量（MSS），当发送窗口中存放的数据达到 MSS 字节数量时，就封装成一个 TCP 报文段发送出去。第二种是采用周期发送方式，发送方维护一个重发定时器，周期性地把当前发送窗口数据封装为不同的 TCP 报文段发送出去，但数据长度不能超过 MSS。第三种是采用推送方式，由发送方的应用进程指明采用 PUSH 操作，发送方将发送缓存中的数据封装成多个不同 TCP 报文段发送出去。

问题 4：在差错控制中，TCP 协议如何确定重发定时器的超时时间？解决思路是什么？

超时重传机制是 TCP 协议中非常重要和复杂的技术之一。TCP 协议每发送一个 TCP 报文段，就对该 TCP 报文段启动一个重发定时器，如果重发定时器超时还没收到确认，就要重传该 TCP 报文段。重发定时器超时时间一般设置为稍大于通信双方的 RTT。

TCP 协议采用一种加权平均往返时间（RTT_S）来计算当前 RTT。当第一次测量到 RTT 样本时，RTT_S 值就取测量到的 RTT 样本值；以后每测量到一个新的 RTT 样本，就按如下公式重新计算一次 RTT_S。

$$新的 RTT_S = (1 - \alpha) \times (旧的 RTT_S) + \alpha \times (新的 RTT 样本)$$

式中，$0 \leq \alpha < 1$；若 α 接近于 0，则表示 RTT 值更新较慢；若 α 接近于 1，则表示 RTT 值更新较快。[RFC2988]标准草案推荐的 α 值为 1/8，即 0.125。

超时重传时间（Retransmission Time-Out，RTO）的计算方法如下。

RTO 应略大于上面得出的 RTT_S。[RFC2988]标准草案建议使用如下公式计算 RTO。

$$RTO = RTT_S + 4 \times RTT_D$$

式中，RTT_D 是 RTT 的偏差的加权平均值。[RFC2988]标准草案建议在第一次测量时，RTT_D 的值取测量到的 RTT 样本值的一半。在以后的测量中，则使用下式计算 RTT_D。

$$新的 RTT_D = (1 - \beta) \times (旧的 RTT_D) + \beta \times |新的 RTT_S - 新的 RTT 样本|$$

式中，β 是小于 1 的系数，其推荐值是 1/4，即 0.25。

问题 5：TCP 协议在计算重发定时器超时时间时存在什么问题？是否可以进一步改进？

在图 9-12 中，TCP 报文段 1 没有收到相应的确认，发送方重传 TCP 报文段 1（标识为 TCP 报文段 2）后，收到了 ACK 应答。如何判定此确认报文段是对应原来 TCP 报文段 1 的确认，还是对重传的 TCP 报文段 2 的确认呢？这直接影响本次 RTT 样本的计算，因此会产生二义性。

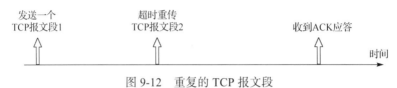

图 9-12　重复的 TCP 报文段

为了解决以上问题，可以采用 Karn 算法。在计算 RTT 样本时，只要 TCP 报文段重传了，就不再采用该 TCP 报文段（包括重发的 TCP 报文段）往返时间样本。这样得出的 RTT_S 和 RTO 就比较准确。但这样会产生两个新问题，具体如下。

（1）由于网络发生了拥塞，因此 TCP 报文段和确认应答在网络上的传输延迟时间增加了。

（2）如果重发定时器超时后还没有收到 ACK 应答，就需要对该 TCP 报文段进行重传；根据 Karn 算法，不需要考虑该 TCP 报文段的 RTT 样本，因此 RTO 时间无法更新。

如何将网络拥塞对 RTO 的影响考虑进去呢？可采用一种修正的 Karn 算法，即 TCP 报文段每重传一次，将 RTO 按照下式增大一些。

$$新的 RTO = \gamma \times (旧的 RTO)$$

式中，系数 γ 的典型值是 2 。当不再发生 TCP 报文段重传时，才根据 TCP 报文段的往返时间更新 RTT_S 和 RTO 的数值。实践证明，这种策略较为合理。

4）选择确认机制（Selective ACK，SACK）

当接收方无差错的收到了和前面字节流不连续的多个字节块，而且这些字节块的序号都在接收窗口之内时，接收方应如何处理？方法一是采用后退 N 帧协议，按乱序处理，丢弃或缓存，但重发时丢失的 TCP 报文段及之后的数据全部重发，TCP 协议一般支持这种方法；方法二是采用选择重发协议，接收方先收下这些不连续的字节块，只重发丢失的 TCP 报文段；方法三是采用选择确认机制，接收方先收下这些不连续的字节块，但要想办法将连续字节块数据准确地告诉发送方，使发送方不需要等重发定时器超时，仅把不连续的字节块重发即可，如图 9-13 所示。

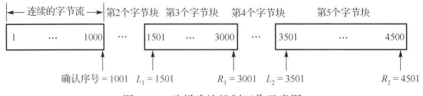

图 9-13　选择确认机制工作示意图

在图 9-13 中，每一个连续的字节块都有两个边界：左边界和右边界，图中用 4 个指针标记这些边界。第 3 个字节块的左边界 L_1 = 1501，但右边界 R_1 = 3001。左边界指出字节块的第 1 个字节的序号，但右边减 1 才是字节块的最后一个字节的序号。第 5 个字节块的左边界 L_2 = 3501，而右边界 R_2 = 4501。[RFC2018]标准草案规定：如果使用选择确认机制，则在建立 TCP 连接时，TCP 首部选项字段中要加上"允许 SACK"选项，而且双方必须都事先商定好。此时，原来 TCP 首部中的"确认序号"字段的用法仍然不变，只是以后在 TCP 报文段的首部（选项字段）中增加了"允许 SACK"选项，以便报告收到的连续字节块的边界（左右边界各 4 字节）。由于首部选项的长度最多只有 40 字节，因此需要一个字节指明要采用的 SACK 选项，另一字节指明该选项实际占用的字节数。指明一个连续字节块需要 8 字节，因此在选项中最多只能指明 4 个连续字节块的边界信息（共 34 字节）。[RFC2018]标准草案没有指明发送方应该如何响应 SACK，因此大多数 TCP 协议的实现还是要采用后退 N 帧协议。

6) 信道利用率

信道利用率（U）（见图 9-14）是指在一次数据传输中，发送数据的时间（T_D）与总时间的比值，其计算公式如下。

$$U = \frac{T_D}{T_D + \text{RTT} + T_A}$$

在信道利用率计算公式中，T_A 表示 ACK 应答的发送时间。假设 A、B 之间的信道距离为 1200km，RTT=200ms，数据长度为 1200 比特，发送速率为 1MB/s，如果忽略处理时间和 T_A（一般 T_A 远远小于 T_D），则 U=5.66%。由此可见，如果流量控制采用停止-等待协议，则 94%以上的时间信道处于空闲状态；如果发送速率提高到 10MB/s，则 U=0.0571%，这时近 99.94%的时间信道处于空闲状态。

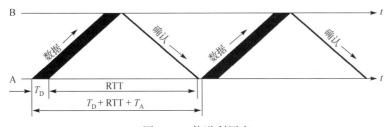

图 9-14 信道利用率

7) 拥塞控制机制

拥塞控制的目的是防止有过多的数据输入网络而造成网络中的路由器（或链路）过载。拥塞控制的原则是网络能够承受现有的网络负荷，所以需要提前进行拥塞控制。拥塞控制是一个全局性过程，涉及网络中所有的主机、网络交换设备（与该路由器通信相关的所有设备）。在 TCP 协议通信过程中，如果发送方重定时器超时后还没有收到 ACK 应答，则认为发生了网络拥塞。发送方通过降低发送速率的方式来解决，但发送方无法知道发生网络拥塞的具体情况，如如何发生拥塞、引起拥塞的具体原因是什么。在实际网络中，随着负载的增加，网络吞吐量的增长率反而降低，特别是当网络发生拥塞时，如果再增加负载到一个数值，则网络吞吐量反而降为 0，如图 9-15 所示。

拥塞控制是一个动态的过程，当前网络朝着高速化方向发展，很容易出现网络交换设备接收缓存不够大（或 CPU 过载）而造成分组丢失（或来不及处理）的情况，这些都是网络发生拥塞的征兆。在许多情况下，可能由于考虑不周，拥塞控制本身会成为引起网络性能恶化甚至发生死锁的原因，这点应特别引起重视。

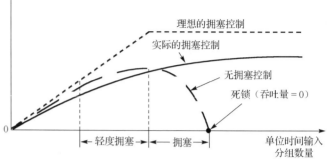

图 9-15　拥塞控制作用示意图

拥塞控制分为开环控制和闭环控制两种方式。开环控制是指在设计网络时事先将发生拥塞的相关因素考虑周到，制定相应的拥塞控制策略，并静态配置到网络交换设备中，以避免拥塞现象的发生；闭环控制是基于反馈概念，一般采用以下三大关键技术。

（1）检测技术：依据拥塞发生时的性能指标，检测何处由于什么原因发生了网络拥塞，一般所有网络交换设备均参与检测过程。

（2）反馈技术：检测点检测到发生网络拥塞后，需要将此信息报告给控制节点，一般可以采用显式反馈和隐式反馈两种方式。

（3）调整技术：控制节点按照一定策略缓解当前的网络拥塞情况，所有网络交换设备和端节点都有可能参与拥塞控制。

拥塞检测基准参数主要有接收队列平均长度、接收缓存缺少空间而丢弃的分组数量、网络中重发分组个数、平均分组延迟、分组延迟标准差。如果上述某个参数超过某门限值，则可怀疑发生了网络拥塞。

拥塞控制和流量控制是两个不同的概念。流量控制在传输层的研究对象为接收方的接收缓存空间，目的是防止接收方的接收缓存溢出，而数据链路层流量控制主要研究如何确保相邻两个网络节点之间接收缓存不溢出的问题。流量控制的目的是抑制发送方发送数据的速率，使接收方来得及从接收缓存区中将数据取走，不让数据从接收缓存区溢出。

1999 年，[RFC2581]标准草案定义了传输层 4 种拥塞控制方法：慢启动（slow start）、拥塞避免(congestion avoid）、快速重传（fast retransmit）和快速恢复（fast recovery）。此后，[RFC2582]、[RFC3390]标准草案对以上方法进行了改进。在介绍传输层拥塞控制前，首先提出以下两个假设。

① 数据是单向传输的，接收方可以发送 ACK 应答。

② 接收方有足够大的接收缓存区，暂时不考虑流量控制，因此发送窗口主要由网络拥塞程度（拥塞窗口）决定。

当网络发生拥塞时，路由器接收缓存溢出，根据拥塞控制策略，路由器需要丢弃分组；由于发送方不可能收到被丢弃分组的对应 TCP 报文段的确认，因此重发定时器超时，需要重发该 TCP 报文段。此时，发送方就认为网络发生了拥塞。实际上，发送方没有收到确认的原因很多，可能是由 TCP 报文段在网络传输过程中出现了差错，接收方直接丢弃引起的，但这时网络并没有发生拥塞，或者确认出现了差错。现在的线路大部分采用光纤，所以发生此类情况的概率很小。

（1）慢启动法。慢启动法的基本思想是，当发送方主机刚开始发送 TCP 报文段时，由于不知道网络的负载情况，可采用试探的方法，由小到大增加拥塞窗口大小（因为不考虑流量控制，所以拥塞窗口大小=发送窗口大小），因此发送方在通信开始时，将拥塞窗口 cwnd 置为1，即设置为一个最大报文段的数值（MSS），第一次只能发送一个 TCP 报文段；发送方每收到一个对新的 TCP 报文段的确认，便将拥塞窗口 cwnd 加 1，即多发送一个 TCP 报文段。实践证明，用这样的方法逐步增大发送方拥塞窗口 cwnd，可使分组输入到网络的速率更加合理。为方便起见，用 TCP 报文段个数作为拥塞窗口大小的单位（实际上是采用字节）。

传输轮次是指 TCP 协议把拥塞窗口 cwnd 内的 TCP 报文段连续发送出去，并将收到的已发送的最后一个字节的确认作为一个轮次传输。使用慢启动方法后，每经过一个传输轮次，拥塞窗口 cwnd 就加倍，即按照指数规律增加。慢启动法中的"慢"并不是指拥塞窗口 cwnd 增长速率慢，而是指当 TCP 协议开始发送报文段时，拥塞窗口 cwnd=1，即只能发送一个 TCP 报文段对网络拥塞情况进行试探，然后逐渐增加拥塞窗口 cwnd，这对防止或避免拥塞是有利的。如果在不知道网络是否存在拥塞的情况下，一下子发送了大量的 TCP 报文段，这是比较危险的，可能会加剧拥塞。

为了防止拥塞窗口 cwnd 增长过大引起网络拥塞，发送方设置了一个慢启动门限 ssthresh，用法如下。

当 cwnd < ssthresh 时，使用慢启动方法；当 cwnd > ssthresh 时，停止使用慢启动方法而改用拥塞避免方法；当 cwnd = ssthresh 时，既可使用慢启动方法，也可使用拥塞避免方法。

（2）拥塞避免法。拥塞避免法是指尽可能让拥塞窗口 cwnd 增长缓慢，即每经过一个传输轮次，便把发送方的拥塞窗口 cwnd 加 1，而慢启动法是加倍。由此可见，拥塞避免法使得拥塞窗口 cwnd 按线性规律缓慢增长。当网络出现拥塞时，无论是在慢启动阶段还是在拥塞避免阶段，只要发送方判断出网络出现拥塞（TCP 协议的判断依据是重发定时器超时），就要把慢启动门限 ssthresh 设置为网络出现拥塞时的发送方窗口值（等于拥塞窗口 cwnd）的一半（不能小于 2），然后把拥塞窗口 cwnd 重新置为 1。由于在整个过程中发送窗口等于拥塞窗口，因此发送窗口的大小也等于 1，最后再执行慢启动方法。这样做的目的是迅速减少主机发送到网络中的 TCP 报文段数量，使发生拥塞的路由器有足够的时间把接收队列中累积的分组处理完毕，如图 9-16 所示。

在图 9-16 中，当建立 TCP 连接初始化时，发送方将拥塞窗口 cwnd 置为 1（图中的窗口单位不使用字节而使用报文段）。慢启动门限的初始值设置为 16 个 TCP 报文段（建立连接初始化时协商），即 ssthresh = 16。发送方的发送窗口不能超过拥塞窗口 cwnd 和接收方的接收窗口有效大小 rwnd 中的最小值。假设接收窗口足够大，则实际发送窗口 cwnd 的数值等于拥塞窗口 cwnd 的数值。在执行慢启动方法时，拥塞窗口 cwnd 的初始值为 1，发送第一个报文段 M_0。发送方每收到一个确认，拥塞窗口 cwnd 加 1，于是发送方可以接着发送 M_1 和 M_2 报文段，接收方共发回两个确认。发送方每收到一个对新报文段的确认，就把发送方的拥塞窗口 cwnd 加 1。现在拥塞窗口 cwnd 从 2 增大到 4，并可接着发送后面的 4 个报文段。发送方每收到一个对新报文段的确认，就把发送方的拥塞窗口 cwnd 加 1，因此拥塞窗口 cwnd 随着传输轮次的增加按指数规律增长。当拥塞窗口 cwnd 增长到慢启动门限值时（当拥塞窗口 cwnd = 16 时），就改用拥塞避免方法，拥塞窗口 cwnd 按线性规律增长。假定拥塞窗口 cwnd 增长到 24，重发定时器超时，表明发生网络拥塞，这时 ssthresh 和拥塞窗口 cwnd 如何变化？拥塞窗口 cwnd 重新置为 1，ssthresh 变为 12[当发送拥塞时，发送窗口（拥塞窗口）的数值为 24 的一半]，并使用慢启动方法（拥塞窗口 cwnd 按指数增加）。当拥塞窗口 cwnd = 12 时，改用拥塞避免方法，拥塞窗口 cwnd 按线性规律增长，每经过一个传输轮次就增加一个 MSS 大小的 TCP 报文段，依次类推。

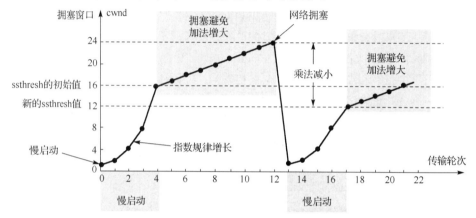

图 9-16 慢启动和拥塞避免方法示意图

拥塞避免法并非能够完全避免拥塞。拥塞避免是指在拥塞避免阶段把拥塞窗口控制为按线性规律增长，使网络较不容易出现拥塞或缓解拥塞程度。乘法减小是指不论在慢启动阶段还是在拥塞避免阶段，只要出现一次重发定时器超时（或出现一次网络拥塞），就把 ssthresh 设置为当前拥塞窗口值的一半，其目的是当网络频繁出现拥塞时，ssthresh 值会下降得很快，以大大减少输入到网络中的 TCP 报文段的数量。加法增大是指在执行拥塞避免法后的一个传输轮次中，发送方收到对所有 TCP 报文段的确认后，就把拥塞窗口 cwnd 增加一个 MSS 大小的 TCP 报文段，使拥塞窗口 cwnd 缓慢增大，以防网络过早出现拥塞。AIMD 称为加法增加乘法减小，后来有人对 AIMD 进行了改进，但现在 TCP 协议使用最广泛的还是 AIMD。慢启动法和拥塞避免法是 1988 年提出的拥塞控制方法，1990 年又增加了两个新的拥塞控制方法，即快速重传法和快速恢复法。

（3）快速重传法。快速重传法首先要求接收方每收到一个失序 TCP 报文段后就立即发出重复确认，让发送方及早知道有报文段没有到达接收方。发送方只要连续收到 3 个重复确认，就应当立即重传对方尚未收到的报文段，不用等到重发定时器超时。快速重传并非取消重发定时器，而是在某些情况下可以更早地重传丢失（或延迟）的 TCP 报文段，如图 9-17 所示。

（4）快速恢复法。快速恢复法一般与快速重传法配合使用（见图 9-18）。当发送方收到连续 3 个重复确认时，就执行乘法减小，即把 ssthresh 减小为当前拥塞窗口 cwnd 的一半，但接下去不执行慢启动法。现在发送方认为网络很可能没有发生拥塞，因为发送方能收到 3 个重复确认，因此不执行慢启动法，拥塞窗口 cwnd 不置为 1，而是将 ssthresh 减小为当前拥塞窗口 cwnd

的一半，同时将拥塞窗口设置为新的 ssthresh 的数值，然后开始执行拥塞避免法（加法增大），使拥塞窗口缓慢地按线性规律增大。综上所述，当连续收到 3 个重复确认时，执行拥塞避免法。

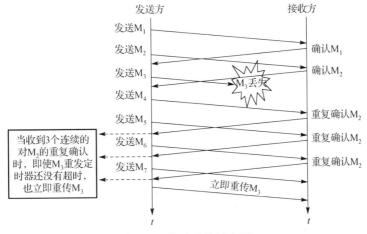

图 9-17　快速重传示意图

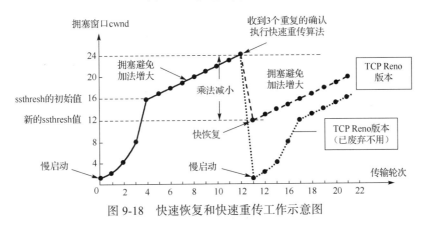

图 9-18　快速恢复和快速重传工作示意图

如果接收方接收缓存有足够大的空间，那么是不用考虑流量控制的。在实际 TCP 协议实现中，由于接收方接收缓存空间有限，因此在考虑拥塞控制的同时，也要考虑流量控制。发送窗口的大小同时受拥塞窗口 cwnd 和接收窗口有效大小 rwnd 的限制，应按以下公式确定。

$$发送窗口的上限值 = \text{Min}\,[rwnd, cwnd]$$

当 rwnd < cwnd 时，由接收窗口控制发送窗口的最大值（流量控制起主要作用）；当 cwnd < rwnd 时，由拥塞窗口控制发送窗口的最大值（拥塞控制起主要作用）。当网络发生拥塞后，路由器一般通过载荷脱落机制来缓解拥塞，其中，用到的随机早期检测（Random Early Detection，RED）技术将路由器的接收缓存队列分为 3 个区域：排队区域、以概率 p 丢弃区域及丢弃区域，如图 9-19 所示。采用平均队列长度 L_{av} 是否大于门限 TH_{min} 作为判断网络是否发生拥塞的依据，当 $L_{av} < TH_{min}$ 时，说明网络没有发生拥塞；当 $L_{av} > TH_{max}$ 时，说明网络有发生拥塞的可能；当 $TH_{min} < L_{av}$ 时，说明网络已经发生了拥塞。

在图 9-20 中，当 $L_{av} < TH_{min}$ 时，丢弃概率 $p = 0$；当 $L_{av} > TH_{max}$ 时，丢弃概率 $p = 1$；当 $TH_{min} < L_{av} < TH_{max}$ 时，$0 < p < 1$。丢弃概率 p 按线性规律变化，变化范围为 $0 \sim p_{max}$。

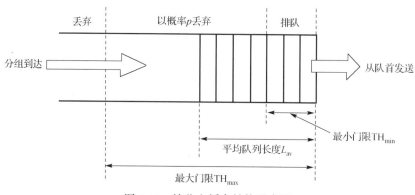

图 9-19 接收方缓存结构示意图

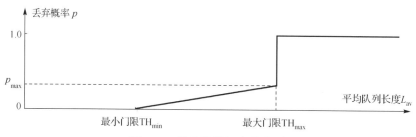

图 9-20 接收数据概率分布图

路由器之所以采用平均队列长度,而不是采用瞬时队列长度作为判断网络是否发生拥塞的依据,是因为采用瞬时队列长度可能会产生误判,或者说误判的概率更大,如图 9-21 所示。

在图 9-21 中,当瞬时队列长度出现尖峰时,路由器可能会判断为网络发生拥塞,但实际上,路由器缓存还没有溢出,如果下一时间段网络负载不断减轻,则在这种情况下,网络拥塞现象可能并没有发生。

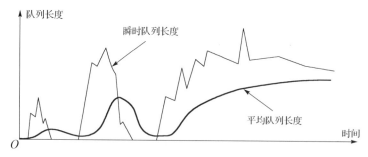

图 9-21 瞬时队列长度和平均队列长度的区别

8)TCP 连接管理

TCP 协议为端到端不同应用进程提供可靠通信服务。TCP 协议通信经过 3 个阶段:建立 TCP 连接、可靠数据传输、释放 TCP 连接。TCP 连接管理可以使连接的建立和释放正常进行,TCP 协议在数据传输阶段维护连接处于正常状态。TCP 连接的建立、可靠数据传输和释放 TCP 连接都是采用客户/服务器方式,主动发起连接请求的应用进程叫作客户,被动等待连接请求的应用进程叫作服务器。在建立连接的过程中要解决以下 3 个问题。

(1)要使一方能够知道对方应用进程的存在。

（2）允许双方协商一些参数（如最大报文段长度 MSS、最大窗口大小与初始序号、是否采用 SACK 及其他服务质量参数等）。

（3）对网络资源（如缓存、各种定时器、连接状态表）进行分配和初始化。TCP 连接表登记了每个连接的信息，除本地和远程的 IP 地址和端口号之外，还记录了每一个连接所处的状态，如表 9-1 所示。

表 9-1　TCP 连接表示例

	本地 IP 地址	本地端口号	远程 IP 地址	远程端口号	连接状态
连接 1					
连接 2					
⋮					
连接 n					

TCP 协议在建立 TCP 连接时的状态变化如图 9-22 所示。

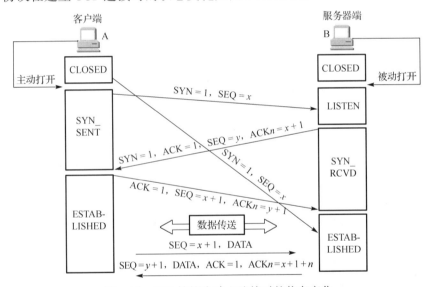

图 9-22　TCP 协议在建立连接时的状态变化

问题：TCP 协议为什么通过 3 次握手建立 TCP 连接？为什么不是通过两次握手建立 TCP 连接？

答案：TCP 协议通过 3 次握手建立 TCP 连接是为了防止将"已延迟 TCP 连接请求"发送给对方。在图 9-23 中，假设主机 A 发出 TCP 连接请求 1，但未收到确认，则主机 A 重传 TCP 连接请求 2，后来收到确认，此时成功建立连接；数据发送完毕后，释放 TCP 连接。在整个过程中，主机 A 发送了两个连接请求报文，其中一个是有效的，暂且认为第二个 TCP 连接请求也是有效的。

在图 9-23 中，如果客户端和服务器端之间采用两次握手而不是 3 次握手建立 TCP 连接，那么此时会出现以下两种情况。

情况一：由于客户端与服务器端之间通信已经结束，因此客户端和服务器端已经释放连接（释放网络资源），但因为第一个连接请求无法到达服务器端，所以新的连接无法在客户端与服务器端之间建立，这种情况没有问题。情况二：客户端已经关机或对应的客户进程已经关闭，因此客户端无法对延时收到的 ACK 应答进行处理（主机不可达或端口不可达），而服务器端以为连接已经建立，所以分配了网络资源，等待客户端发送 TCP 报文段，如图 9-24 所示。$\langle SYN=1，SEQ=x \rangle$ 对应的连接请求有可能错误地建立客户端到服务器端的 TCP 连接上，这样容易造成网络资源浪费，甚至给数据传输安全带来风险。

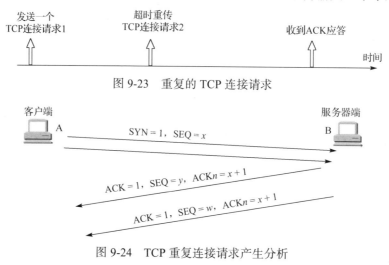

图 9-23 重复的 TCP 连接请求

图 9-24 TCP 重复连接请求产生分析

如果采用 3 次握手建立 TCP 连接，则可通过以下 3 种情况进行问题分析。

情况一：假设客户端与服务器端之间采用 3 次握手建立 TCP 连接，当客户端与服务器端通信时，如果正常连接刚建立好，服务器端就收到客户端发送的延迟第一个连接请求，所以延迟连接请求序号不在接收窗口内，因此该连接请求会被丢弃；如果服务器端在 ESTABLISHED 状态前收到延迟连接请求，由于接收窗口序号重叠，服务器端会认为该连接请求是一个重复 TCP 报文段而丢弃；如果服务器端在 ESTABLISHED 状态后收到延迟连接请求，由于接收窗口从 $x+1$ 开始，而该 TCP 报文段序号为 x，因此延迟连接请求的 TCP 报文段序号不在接收窗口内而丢弃。

在以上 3 种情况下，由于在 TCP 连接建立过程中，3 次握手是一个原子操作，因此延迟连接请求无法产生错误的 TCP 连接，如图 9-25 所示。

情况二：假设客户端与服务器端之间采用 3 次握手建立 TCP 连接，那么客户端与服务器端在通信结束时，TCP 连接已释放、客户端对应的客户进程已经停止运行，此时第一个连接请求到达服务器端。由于客户端的客户进程已经关闭，服务器端发送的 ACK 应答和反向连接请求到达客户端后，TCP 协议无法找到端口号对应的接收队列，因此客户端的 TCP 协议实体会丢弃该 ACK 应答，同时给服务器端发送"端口不可达"ICMP 差错报告报文。由于客户端不会给服务器端发送 ACK 应答，所以错误的 TCP 连接无法建立，如图 9-26。

情况三：假设客户端与服务器端之间采用 3 次握手建立 TCP 连接，那么客户端与服务器端在通信结束时，TCP 连接已经释放，并且客户端已经关机，此时第一个连接请求到达服务器端。由于客户端关机，服务器端发送的 ACK 应答和反向连接请求在最后一个路由器不会转发给客户端（因为 ARP 协议），因此该路由器丢弃此连接请求，并发送"目的不可达（主机不可达）"ICMP 差错报告报文给服务器端。客户端不会给服务器端发送 ACK 应答。由于 3 次握手是原子操作，客户端没有发送应答给服务器端，因此客户端与服务器端之间的 TCP 连接无法建立。

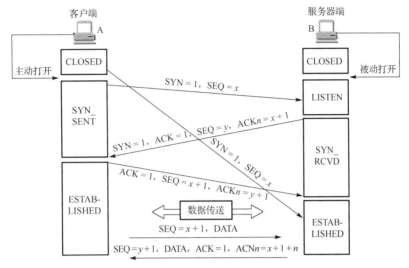

图 9-25　服务器端在 ESTABLISHED 状态下收到延迟连接请求

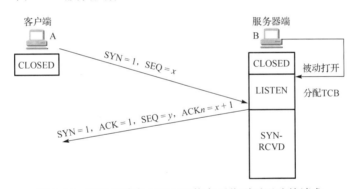

图 9-26　服务器端在 LISTEN 状态下收到延迟连接请求

TCP 协议采用 4 次挥手释放已经建立的 TCP 连接。在图 9-27 中，TCP 协议在释放连接时必须经过 2MSL 时间后才能真正释放响应的网络资源（MSL 表示最大报文段寿命，一般情况下 MSL = 2min，所以 2MSL = 4min）。造成这样的原因如下。

（1）为了确保在 4 次挥手释放 TCP 连接的过程中最后一个握手信号 ACK 应答到达接收方（服务器端），假设服务器端没有收到对方发送的最后一个 ACK 应答报文，此时造成服务器端的释放连接请求 FIN 报文重发定时器超时，服务器端会重新发送释放连接请求报文，此时发送方（客户端）有机会重新发送 ACK 应答，并重新启动 2MSL 计时器。

（2）由于这些报文段序号不在服务器端的接收窗口序号内，为了使延定的连接请求 TCP 报文段或延迟的 TCP 报文段在 2MSL 时间内有机会发送给服务器端，因此服务器端会丢弃这些报文段序号，目的是让本连接中延迟的 TCP 报文段在本连接持续时间内从网络中消失，不要出现在新的连接中对通信产生影响。

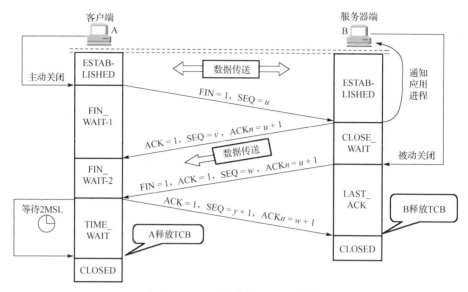

图 9-27 4 次挥手释放 TCP 连接

TCP 有限状态机如图 9-28 所示。在图 9-28 中，每个方框都表示 TCP 连接可能具有的状态，每个方框中的大写英文字符串是 TCP 协议使用的 TCP 连接状态名；状态之间带箭头的连线表示可能发生的状态变迁；连线旁边的字表明引起状态变迁的原因（发生了什么事件）。在图 9-28 中，有 3 类不同的带箭头的连线，粗实线箭头表示对客户端的正常状态变迁，虚线箭头表示对服务器端的正常状态变迁，细实线箭头表示异常变迁。这是设计状态检测防火墙的理论依据。

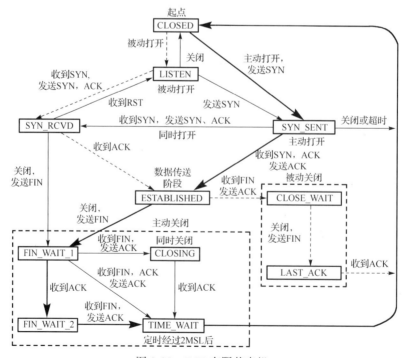

图 9-28 TCP 有限状态机

9.3 TCP 协议发送数据段

TCP 协议发送数据段源代码如下。

```c
#include "iostream.h"
#include "winsock2.h"
#include "ws2tcpip.h"

#pragma comment(lib, "ws2_32.lib")
//定义IP分组首部数据结构
Typedef struct tagIP_Header
{
   Union
   {
      Unsigned char verion;              //版本号（字节的前4比特）
      Unsigned char HeadLength;   //首部长度（字节的后4比特，以4字节为计算单位）
   }
   Unsigned char tos;                    //服务类型
   Unsigned short totallength;           //分组总长度（首部+数据）
   Unsigned short identifier;            //标识
   Union
   {
      Unsigned short flags;              //标志（前3比特）
      Unsigned short fragementoffset;    //片偏移（后13比特）
   }
   Unsigned char ttl;                    //TTL
   Unsigned char protocol;               //协议类型
   Unsigned short HeaderChecksum;        //首部简单校验和
   Unsigned int srcAddress;              //源IP地址
   Unsigned int dstAddress;              //目的IP地址
}IPHeader, *ptrIPHeader;
//定义TCP伪首部数据结构，在计算TCP首部校验和时用
Typedef struct tagPseTCPHeader
{
   Unsigned int srcAddress;              //源IP地址
   Unsigned int dstAddress;              //目的IP地址
   Unsigned char reserved;               //保留
   unsigned char protocolType;           //协议类型
   Unsigned short TCPTotalLen;           //TCP报文段总长度（首部+数据）
}PseTCPHeader

//定义TCP报文段首部数据结构
Typedef struct tagTCPHeader
{
   Unsigned short srcPort;               //源端口号
```

```
    Unsigned short   dstPort              //目的端口号
    Unsigned int sequence;                //序号
    Unsigned int ackNumber;               //应答序号
    Union
    {
        Unsigned short HeaderLen;    //TCP首部长度（字节的前4比特，以4字节为单位）
        Unsigned short Reserved;     //保留（字节中6比特）
        Unsigned short flags;        //标志位（字中后6比特）
    }节
    Unsigned short winSize;               //接收方有效窗口大小
    Unsigned short checksum;              //校验和（伪首部+首部+数据）
    Unsigned short urgentOffset;          //紧急指针
}TCPHeader, *ptrTCPHeader;

Unsigned short checkBuffer[65535];
Char send_buffer[65535];
Const char tcpData[]={"the tcp packet encapsule and send!"};

//计算校验和函数
Unsigned short CheckSum(unsigned short *ptrBuffer, int iSize)
{
    Unsigned long ResultCheck = 0;
    While (iSize >1)
    {
        ResultCheck += *ptrBuffer++;
        Size -= sizeof(unsigned short);
    }
    If(iSize)
        ResultCheck += *(unsigned char*)ptrBuffer;
    ResultCheck = (ResultCheck>>16) +( ResultCheck&0Xffff);
    ResultCheck += (ResultCheck>>16);
    Return (unsigned short)(~ ResultCheck);
}

Void main(int argc, char* argv[])
{
    //初始化动态链接库
    WSADATA WSAData;
    WSAStartup(MAKEWORD(2,2),&WSAData)
    //建立Raw Socket
    SOCKET RawSocket = socket(AF_INET, SOCK_RAW, TPPROTO_IP)
    //设置发送超时时间
    Int timeout = 1500;
    Setsockopt(RawSocket, SOL_SOCKET, SO_SNDTIMEO, (char *)&timeout,
            sizeof(timeout));
    //设置IP首部控制选项
```

```
Int sendflag =1;
Setsockopt(RawSocket, IPPROTO_IP,IP_HDRINCL, (char*)&sendflag,sizeof(sendflag));

//填充IP报文段首部各字段
IPHeader = {0};
IPHeader.version = (0x04<<4|sizeof(IPHeader)/4);
IPHeader.tos = 0;
IPHeader.totallength = sizeif(IPHeader) + sizeof (TCPHeader) + sizeof(tcp_data);
IPHeader.identifier = 0;
IPHeader.ttl = 64;
IPHeader.protocoltype=IPPROTO_TCP;
IPHeader.HeaderChecksum = 0
IPHeader.srcAddress = inet_addr("192.168.0.1");
IPHeader.dstAddress = inet_addr("192.168.0.2");

//填充TCP伪首部各字段
pseTCPHeader ={0};
pseTCPHeader.srcAddress = IPHeader.srcAddress;
pseTCPHeader.dstAddress = IPHeader.dstAddress;
pseTCPHeader.reserved = 0;
pseTCPHeader.protocolType = IPHeader.protocoltype;
IPHeader. TCPTotalLen = sizeof(TCPHeader) + sizeif(tcpData);
//填充TCP报文段首部各字段
TCPHeader = {0};
TCPHeader .srcPort = 200;
TCPHeader .dstPort = 300;
TCPHeader .sequence = 0;
TCPHeader .ackNumber = 0;
TCPHeader .HeaderLen = ((sizeof(TCPHeader)/4)<<4|0);
TCPHeader .winSize = 8500;
TCPHeader .checksum = 0;
TCPHeader .urgenOffset = 0;

//计算TCP报文段首部简单校验和字段
Memset(checkBuffer, 0, 65535);
Memcpy(checkBuffer, &pseTCPHeader,sizeof(pseTCPHeader));
Memcpu(checkBuffer + sizeif(pseTCPHeader), & TCPHeader,sizeof(TCPHeader));
Memcpu(checkBuffer + sizeif(pseTCPHeader) + sizeif(TCPHeader),
&tcpData,sizeof(tcpData));
TCPHeader .checksum = checksum(checkBuffer, sizeif(pseTCPHeader) +
        sizeif(TCPHeader)
+ sizeof(tcpData));

//构造TCP报文段
Memset(send_buffer, 0, 65535);
Memcpy(send_buffer, &IPHeader, sizeof(IPHeader));
```

```
    Memcpy(send_buffer + sizeof(IPHeader),&TCPHeader, sizeof(TCPHeader));
    Memcpy(send_buffer +sizeOf(IPHeader) +sizeof(TCPHeader),tcpData,sizeof(tcpData));

    //填充Socketaddr_in数据结构
    Socketaddr_in destSocket_addr;
    destSocket_addr.sin_family = AF_INET;
    destSocket_addr.sin_addr.s_addr = inet_addr("192.168.0.2");

    //发送构造好的TCP报文段
    int iSendSize = sizeOf(IPHeader) +sizeof(TCPHeader) + sizeof(tcpData);
    int iResultSize = sendto(RawSocket, send_buffer, iSendSize, 0,
    (struct sockaddr *)& destSocket_addr,sizeof(destSocket_addr));
    Cout<<"send TCP segment is successful!"<<endl;
    Colsesocket(RawSocket);
    WSACleanup();
}
```

9.4 TCP 协议接收数据段

基于 TCP 协议的并发服务器端在设计和实现时的具体工作流程如下。

（1）服务器端主线程首先创建一个侦听套接字。

（2）在侦听套接字上绑定服务器端 IP 地址和端口号。

（3）侦听套接字侦听是否有客户端的 TCP 连接请求到来，如果收到连接请求，则使得侦听套接字处于侦听状态，否则继续侦听。

（4）当客户端请求到来时，由主线程负责创建一个服务器端套接字和服务线程。

（5）将服务套接字作为参数传递到服务线程，主线程返回步骤（4）。

（6）此时由服务线程负责接收客户端的请求，并对该请求进行处理，最后将处理结果反馈给客户端。

（7）服务线程关闭服务器端套接字，并结束服务线程。

基于 TCP 协议的并发服务器端源代码如下。

```
#pragma comment(lib, "ws2_32.lib")
#include "winsock2.h"
#include "process.h"
#define ThreadNum = 10;
Void WorkThread(LPVOID lpParam)
{
    SOCKET sockServer = (SOCKET) lpParam;
    fd_set readset;
    int ret;
    timeval tv;
    char buf[5000];
    While(1)
    {
```

```
        FD_ZERO(&readSet);
        FD_SET(sockserver, &readSet);
        tv.tv_sec =10;
        tv.tv_usec = 0;
        ret = select(0,&readSet, NULL, NULL,&tv);
        if(ret == SOCKET_ERROR|| ret ==0)  //error or timeout
        {
            Printf("select error:%d or timeout\n",WSAGetLastError());
            Break;
        }
        If(FD_ISSET(sockSvr,&readSet))
        {
            Memset(buf,0,5000);
            Ret = recv(sockServer, buf, 5000,0);
            if(ret == SOCKET_ERROR|| ret ==0)  //error or timeout
            {
                Printf("recv error: %d, or peer closed!\n". WSAGetLastError());
                Break;
            }
            Ret = send(sockServer, buf, strlen(buf),0);
            If (ret == SOCKET_ERROR)
            Break;
        }
    }
    Closesocket(sockServer);
}

Int main(int argc, char *argv[])
{
    WSAData wsaData;
    WSAStartup(WINSOCK_VERSION, &wsaData);

    SOCKET sockListen = socket(AF_INET, SOCK_STREAM,0);
    bool bResult = true;
    setsockop(sockListen, SOL_SOCKET, SO_REUSEADDR,(char *)&bResult, sizeof(bResult));
    struct sockaddr_in serverAddr;
    memset(&serverAddr,0,sizeof(serverAddr));
    serverAddr.sin_addr.s_addr = INADDR_ANY;
    serverAddr.sin_family = AF_INET;
    serverAddr.sin_port = htons(8888);
    if(bind(sockListen,(struct sockaddr *)&serverAddr, sizeof(serverAddr))==SOCKET_ERROR)
    {
        Printf("bind error : %d\n",WSAGetLastError());
        Closesocket(sock);
        WSACleanup();
        Return -1;
```

```
    }
    If(listen(sockListen,5) == SOCKET_ERROR)
    {
        Printf("listen error:%d\n",WSAGetLastError());
        WSACleanup();
        Return -1;
    }
    SOCKET sockAccept;
    While(true)
    {
        sockAccept = accept(sockListen,NULL,NULL);
        if(sockAccept == INVALID_SOCKET)
           break;
        else
           _beginthread(WorkThread, 0, (LPVOID) sockAccept);
    }
    Closesocket(sockListen);
    WSACleanup();
    Return 0;
}
```

为了提高 TCP 协议服务器端运行效率，在开发过程中，通常采用在进程初始化时预先创建多个工作线程，即线程池。线程池的一般开发流程如下。

（1）主线程创建侦听套接字。
（2）绑定服务器端 IP 地址和端口号。
（3）调用 listen()函数，使侦听套接字处于侦听状态。
（4）以侦听套接字为参数，创建多个服务线程，主线程进入休眠状态。

服务器端线程池的工作流程如下。

（1）接收客户端连接请求。
（2）Accept()函数返回服务器端套接字。
（3）处理客户端请求。
（4）将处理结果通过服务器端套接字反馈给客户端。
（5）关闭服务器端套接字。
（6）服务线程返回步骤（1）。

基于线程池的 TCP 协议服务器端源代码如下。

```
#pragma  comment(lib, "ws2_32.lib")

#include "winsock2.h"
#include "process.h"
#define ThreadNum = 10;
Void WorkThread(LPVOID lpParam)
{
    SOCKET sockSvrListen = (SOCKET) lpParam;
```

```
SOCKET sockSrv;
fd_set readset;
int ret;
timeval tv;
char buf[5000];

while (true)
{
    sockSrv = accept(sockListen,NULL, NULL);
    if(sockSvr == INVALID_SOCKET)
    {
        printf("accept return error:%d\n", WSAGetLastError());
        continue;
    }
    While(1)
    {
        FD_ZERO(&readSet);
        FD_SET(sockSvr, &readSet);
        tv.tv_sec =10;
        tv.tv_usec = 0;
        ret = select(0,&readSet, NULL, NULL,&tv);
        if(ret == SOCKET_ERROR|| ret ==0)  //error or timeout
        {
            Printf("socket id: %d,select error:%d or timeout\n". sockSvr,
                WSAGetLastError());
            Break;
        }
        If(FD_ISSET(sockSvr,&readSet))
        {
            Memset(buf,0,5000);
            Ret = recv(sockSvr, buf, 5000,0);
            if(ret == SOCKET_ERROR|| ret ==0)  //error or timeout
            {
                Printf("recv error: %d, or peer closed!\n". sockSvr,
                    WSAGetLastError());
                Break;
            }
            Ret = send(sockSvr, buf, strlen(buf),0);
            If (ret == SOCKET_ERROR)
            Break;
        }
    }
    Closesocket(sockSvr);
}
```

```
Int main(int argc, char *argv[])
{
    WSAData wsaData;
    WSAStartup(WINSOCK_VERSION, &wsaData);

    SOCKET sock = socket(AF_INET, SOCK_STREAM,0);
    bool bResult = true;
    setsockop(sock, SOL_SOCKET, SO_REUSEADDR,(char *)&bResult, sizeof(bResult));
    struct sockaddr_in serverAddr;
    memset(&serverAddr,0,sizeof(serverAddr));
    serverAddr.sin_addr.s_addr = INADDR_ANY;
    serverAddr.sin_family = AF_INET;
    serverAddr.sin_port = htons(8888);
    if(bind(sock,(struct sockaddr *)&serverAddr, sizeof(serverAddr))==
            SOCKET_ERROR)
    {
        Printf("bind error : %d\n",WSAGetLastError());
        Closesocket(sock);
        WSACleanup();
        Return -1;
    }
    If(listen(sock,5) == SOCKET_ERROR)
    {
        Printf("listen error:%d\n",WSAGetLastError());
        WSACleanup();
        Return -1;
    }
    For (int I =0; i< ThreadNum; i++)
    _beginthread(WorkThread,0,(LPVOID)sock);

    Sleep(INFINITE);

    Closesocket(sock);
    WSACleanup();
    Return 0;
}
```

第 10 章 应用层协议分析与实践

10.1 引 言

基于 TCP/IP 协议的应用层协议有很多，因为面向网络的应用是多种多样的，所以每一种面向网络的应用都可能对应一种应用层协议。例如，在因特网中，几乎所有的应用系统都有与之对应的应用层协议提供支持，如 HTTP 协议支持 Web 应用、SMTP 协议支持电子邮件应用、Telnet 协议支持远程登录应用、FTP 协议支持文件传输应用、Gopher 协议支持信息检索应用，以及 DNS 协议支持域名系统等。此外，应用层协议还有用于网络安全的安全协议（如 SHTTP 协议）、用于网络管理的网管协议（如 SNMP 协议），以及用于多媒体会议的通信协议等。为支持网络应用的开发，TCP/IP 协议提供了一种套接字网络编程接口，开发者可以通过套接字网络编程接口调用传输层或网络层的服务功能来开发面向特定应用的应用层协议。

Web 是因特网中非常受欢迎的一种多媒体信息服务系统，它基于客户/服务器模式，整个系统由 Web 服务器、浏览器（Browser）和通信协议等三部分组成。其中，通信协议采用超文本传输协议（HTTP 协议）。HTTP 协议是专门为 Web 服务系统设计的应用层协议，能够传送任意类型的超媒体数据对象，以满足 Web 服务器与浏览器之间的多媒体通信需要。

在 Web 服务器上，多媒体信息以网页的形式来发布，网页采用超文本标记语言（Hyper Text Markup Language，HTML）编写。客户使用浏览器连接到指定的 Web 服务器并获取网页后，由浏览器来解释执行网页包含的信息，并显示在客户端主机的屏幕上。在 Web 系统中，使用了一种简单的命名机制：统一资源定位符（Universal Resource Locator，URL）唯一地标识和定位因特网中的资源。例如，URL:http://some.site.edu/somedir/welcome.html。URL 并不仅限于描述 Web 服务器资源，而且可以描述其他服务器资源，如 Telnet、FTP、Gopher、Wais 和 Usenet News 等。

FTP 协议是一种基于 TCP 协议的应用层文件传输协议，也是 UNIX 系统提供的一种重要的网络服务，并广泛用于因特网中。FTP 协议主要用于两个主机之间传输各种文件，它采用客户/服务器模式，FTP 客户程序必须与远程的 FTP 服务器建立连接，在登录认证完成后，才能进行文件传输。通常，用户必须在 FTP 服务器进行注册，即建立用户账号，在拥有了合法的用户名和口令后，才有可能进行有效的 FTP 连接、登录和传输文件。[RFC959]

标准草案详细描述了 FTP 协议功能。在因特网中，存在着成千上万个 FTP 服务器，因此要求用户在每个 FTP 服务器上都建立用户账号是不现实的，也是不可能的。实际上，因特网中的 FTP 服务器采用匿名（anonymous）认证方式，即 FTP 服务器的提供者设置了一个特殊的用户名，如 anonymous，提供给公众使用，任何用户都可以使用这个用户名与提供这种匿名 FTP 服务器的主机建立连接，并共享这个主机对公众开放的资源。匿名 FTP 服务器的用户名是 anonymous，而密码通常是 guest 或使用者的 E-mail 地址。当用户登录到匿名 FTP 服务器后，其工作方式与传统的 FTP 协议相同。通常，出于安全的目的，大多数匿名 FTP 服务器只允许下载文件，而不允许上传文件。也就是说，用户只能从匿名 FTP 服务器复制所需的文件，而不能将文件复制到匿名 FTP 服务器上。此外，匿名 FTP 服务器中的文件还加入了一些保护措施，以确保这些文件不能被修改和删除，同时可以防止计算机病毒的侵入。在因特网中，众多的匿名 FTP 服务器形成一个巨大的文件库，这些文件中包含了各种各样的信息、数据和软件（源程序和可执行程序）。通过 FTP 服务器传输的文件可以是文本文件（ASCII 码），也可以是二进制（binary）文件，用户可以使用 FTP 命令（ASCII 码/binary）来定义文件类型。

10.2　HTTP 协议工作原理

　　Web 最初是由欧洲高能物理研究中心（CERN）开发的，目的是为科研人员共享学术信息提供一种有效的信息服务手段。时隔不久，Web 就作为一种通用的信息发布和服务系统被广泛用于因特网。浏览器软件提供多媒体信息发布服务，这些信息通过网页形式来组织和存储，而网页则采用 HTML 语言来编写。对于浏览器软件来说，除提供响应浏览器的请求发送网页的基本功能之外，还提供了很多其他系统功能，如系统管理、安全管理、编程接口、网页编辑、用户跟踪等。目前，浏览器软件有很多种，这些浏览器软件一般都基于某种操作系统平台并各具特色。例如，Microsoft 公司基于 Windows NT 平台的 Internet Information Server（IIS），Netscape 公司基于 Windows NT 平台和 UNIX 平台的 Enterprise Server，Silicon Ghaphics 公司基于 UNIX 平台的 WebFore Series 及基于 Linux 平台的共享软件 Apache 等。浏览器软件主要用于连接 Web 服务器、解释执行由 HTML 编写的文档，并将执行结果显示在客户机屏幕上，其不仅提供了网页的查找和显示功能，而且提供了网页管理功能及其他客户工具。

　　目前，浏览器软件有很多种，比较著名的浏览器软件有 Microsoft 公司的 Internet Explorer（IE）、Netscape 通信公司的 Navigator 及 Sun 公司的 Hot Java 等。最流行的浏览器软件是 IE 浏览器，Microsoft 公司采取将 IE 浏览器与 Windows 平台一起捆绑销售的策略，使 IE 浏览器拥有众多的用户，占有很大的市场份额。在因特网发展早期，FTP 协议传输文件大约占了网络通信流量的 1/3；1995 年，Web 通信量首次超过了 FTP 协议。Web 是欧洲粒子物理实验室的 Tim Berners-Lee 于 1989 年提出的，如图 10-1 所示。1993 年，第一个图形界面浏览器（Mosaic）研发成功；1995 年，Netscape Navigator 浏览器上市，如图 10-2 所示。

　　万维网的出现是因特网发展中一个非常重要的里程碑，它使因特网由少数计算机专家使用变成普通百姓也可以参与，实现了资源共享，网站数量也按指数规律增长。万维网不是一个特殊的计算机网络,而是一个大规模的、分布式的信息存储方式。采用"链接"的方式,用户可以在因特网上以一个站点访问另一个站点的网页(超媒体信息:文本、视频、音频、图像),用户还可以主动地按需获取丰富的信息。万维网以客户/服务器方式工作,客户程序向服务器程序发出请求,服务器程序向客户程序返回客户所要的万维网文档或处理结果,如图 10-3 所示。浏览器就是在用户计算机上的万维网客户程序。万维网服务器由以下两部分构成。

(1) Web 服务器管理程序:对万维网文档进行管理。

(2) 万维网文档(信息资源):网站(网页+数据库)。

图 10-1　欧洲粒子物理实验室的 Tim Berners-Lee

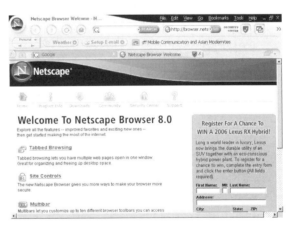

图 10-2　Netscape Navigator 浏览器

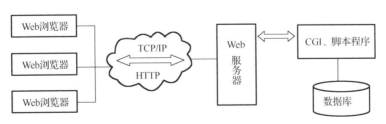

图 10-3　Web 系统工作原理示意图

分布在因特网上的 Web 文档使用统一资源定位符(URL)来标识。一个文档(文件)在因特网围内具有唯一的 URL。万维网客户程序与万维网服务器之间在进行交互时使用的协议为 HTTP 协议。HTTP 协议是一个应用层协议，在传输层使用 TCP 连接进行可靠的通信。万维网文档采用 HTML 编写，浏览器利用自身解析器对文档进行解析，并将解析结果显示在屏幕上；万维网文档中的"链接"采用特殊标识，如鼠标手形、特殊颜色及下划线等。为了方便在万维网上查找信息，用户可使用各种搜索工具（搜索引擎）。

10.2.1 统一资源定位符

URL 可以唯一地标识因特网上任一资源（位置+名称）和访问方法。URL 给资源（位置+名称）提供一种抽象的标记和识别方法，用该方法可对资源进行定位。只要能够对资源进行定位，系统就可以对资源进行各种操作，如存取、更新、替换和查找其属性。URL 相当于一个文件名在网络范围的扩展，其由以冒号隔开的两大部分组成，并且 URL 中的字符对大写或小写没有要求。URL 的一般形式如图 10-4 所示。

<协议>://<主机>:<端口>/<路径+名称>

图 10-4 URL 的一般形式

在 URL 中，若省略文件的<路径+名称>项，则 URL 指到因特网上的某个主页。例如，http://www.nwpu.edu.cn:80/index.html（.htm）可以略写为 http://www.nwpu.edu.cn 或 www.nwpu.edu.cn。

10.2.2 HTTP 1.0 协议的主要特点

1）HTTP 1.0 协议是无状态协议

HTTP 协议是一种面向事务的客户服务器通信协议，HTTP 1.0 协议是无状态的协议。当同一用户第二次访问同一服务器的同一页面时，服务器的应答和第一次应答一样，服务器不记录谁曾访问过，又访问过几次，以及访问了哪些内容。这样设计的目的是简化了服务器，使服务器容易支持大量的并发 HTTP 请求。HTTP 协议本身是无连接的，在传输层选用了面向连接的 TCP 协议，目的是保证数据传输的可靠性。在 1997 年以前，Web 系统一般采用 HTTP 1.0 协议（[RFC1945]标准草案），之后升级为 HTTP 1.1 协议（[RFC2616]标准草案）。

请求一个万维网文档所需的时间，主要包括通过 3 次握手建立 TCP 连接的时间和传输 HTTP 响应报文的时间。为了节省时间，系统在发送第 3 次握手应答信号时，捎带客户对 Web 文档的请求，如图 10-5 所示，因此用户获得一个万维网文档的时间为 2RTT+文档发送时间。

2）HTTP 1.0 协议采用一次一连接

一次一连接是指 HTTP 1.0 协议在工作时，如果客户端请求一个数据对象，则客户端与服务器端之间先通过 3 次握手建立一个 TCP 连接，然后发送 HTTP 请求给服务器端，之后服务器端对客户端请求进行处理并发送 HTTP 应答给客户端，客户端接收后解析显示，并通过 4 次挥手释放该连接。如果客户端需要再次请求一个数据对象，则需要重复以上操作。当一个网页文档中有 9 个图片链接对象和文本信息时，下载文档文本信息需要 1 个 TCP 连接，下载 9 个图片需要 9 个 TCP 连接。由于浏览器一般允许同时建立 5~10 个 TCP 连接，因此可实现并行传输，这在一定意义上缩短了响应时间，但客户端和服务器端消耗了大量的网络资源。

HTTP 1.1 协议解决了上述资源消耗的问题，其采用持续连接方式（persistent connection）。在浏览器与服务器通信之前，首先建立好一个 TCP 连接，在通信阶段浏览器可在该连接上依次发送多个 HTTP 请求报文，服务器也可以依次连续发送多个 HTTP 响应报文；在此通信会话期间，该 TCP 连接一直保持，实现了在一个 TCP 连接上可连续传输一个文档的多个链接，条件是每个链接的对象，如图片、文本、视/音频等信息必须在同一个服务器上。目前，一些流行的浏览器（如 IE 6.0）的默认设置就是使用 HTTP 1.1 协议。

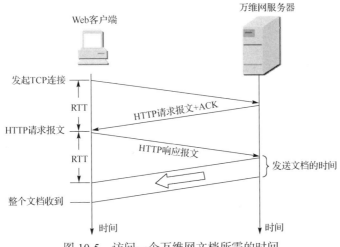

图 10-5　访问一个万维网文档所需的时间

持续连接采用两种工作方式，第一种为非流水线方式，类似停止-等待协议工作原理。在一个 TCP 连接上，客户端在收到前一个 HTTP 响应报文后才能发出下一个 HTTP 请求报文。在这种通信方式中，服务器端在发送完一个数据对象后，其 TCP 连接有可能处于空闲状态，因此浪费了服务器端资源。第二种为流水线方式，类似连续 ARQ 协议工作原理。在一个 TCP 连接上，客户端在收到 HTTP 响应报文之前，能够接着连续发送多个新的 HTTP 请求报文，一个接一个的 HTTP 请求报文到达服务器端后，服务器端就可以连续回送 HTTP 响应报文。用户打开客户端浏览器，单击"工具"→"Internet 选项"选项，打开"Internet 选项"对话框，单击"高级"选项卡，如图 10-6 所示。如果用户想在客户端浏览器和 Web 服务器之间采用 HTTP 1.1 协议通信，则需要勾选"使用 HTTP 1.1"复选框，否则标识采用 HTTP 1.0 协议。

图 10-6　HTTP 协议版本号设置

10.2.3 Web 代理服务器

Web 代理服务器又称为万维网高速缓存。Web 代理服务器把最近的一些 HTTP 请求报文和响应报文暂存在本地的磁盘缓存中,当新的请求报文到达时,如果 Web 代理服务器发现该请求报文与暂存的某一请求报文相同,则返回暂存的相应响应报文,不需要按 URL 的地址再去因特网访问该资源。Web 代理服务器可部署在客户端、服务器端或中间系统上工作。

在图 10-7 中,当客户端浏览器访问因特网的服务器时,要先与校园网的 Web 代理服务器建立 TCP 连接,并向 Web 代理服务器发出 HTTP 请求报文;若 Web 代理服务器已经存放了请求的对象,则将此对象放入 HTTP 响应报文中返回给客户端浏览器,否则 Web 代理服务器就代表发出请求的客户端浏览器,并与因特网上的 Web 服务器建立 TCP 连接,然后发送 HTTP 请求报文,最后 Web 服务器对该请求进行处理,并将请求的对象放在 HTTP 响应报文中返回给校园网的 Web 代理服务器。Web 代理服务器收到此对象(HTTP 应答报文)后,先复制在本地存储器中,以方便今后使用,然后将该对象放在 HTTP 响应报文中,通过已建立的 TCP 连接返回给请求该对象的客户端浏览器。用户可以打开客户端浏览器,单击"工具"→"Internet 选项"选项,在"局域网(LAN)设置"对话框中单击"高级"按钮,然后在打开的"代理服务器设置"对话框中设置 Web 代理服务器的相关信息,如图 10-8 和图 10-9 所示。

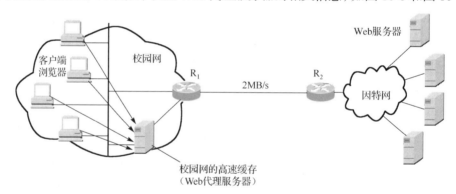

图 10-7 Web 代理服务器工作示意图

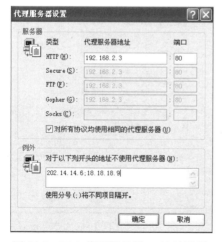

图 10-8 Web 代理服务器 IP 地址设置

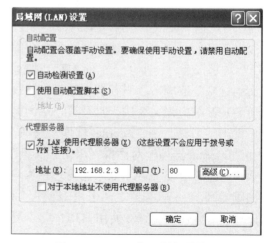

图 10-9 Web 代理服务器设置

如果用户想在客户端浏览器和 Web 服务器之间采用 Web 代理服务器作为中介进行通信，并与 Web 代理服务器之间采用 HTTP 1.1 协议通信，则需要打开客户端浏览器，单击"工具"→"Internet 选项"选项，打开"Internet 选项"对话框，单击"高级"选项卡，并勾选"通过代理连接使用 HTTP 1.1"复选框，如图 10-10 所示。

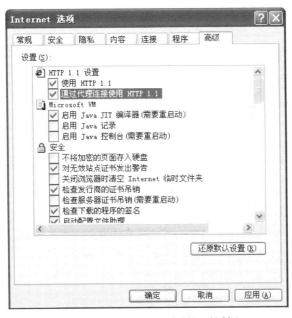

图 10-10　"Internet 选项"对话框

10.2.4　HTTP 报文的语法和语义

HTTP 协议提供了以下两类报文（语义）。

HTTP 请求报文：从客户端端浏览器向服务器端发送 HTTP 请求报文。

HTTP 响应报文：从服务器端反馈到客户端浏览器的 HTTP 响应报文。

HTTP 请求报文由 3 部分组成：开始行、首部行和实体主体，如图 10-11 所示。

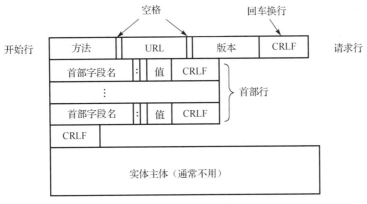

图 10-11　HTTP 请求报文格式

在 HTTP 请求报文中，开始行就是请求行，一般实体主体不用。开始行中的"方法"是面向对象技术中使用的专业名词，这里表示对请求的对象进行的操作类型，因此这些方法实际上也是通过一些操作命令来实现的，如表 10-1 所示。

表 10-1 HTTP 协议支持的方法

方法（操作）	意义
OPTION	请求一些选项的信息
GET	请求读取由 URL 标识的信息
HEAD	请求读取出 URL 标识的信息的首部
POST	给服务器添加信息（如注释）
PUT	在指明的 URL 下存储一个文档
DELETE	删除指明的 URL 标识的资源
TRACE	用来进行环回测试的请求报文
CONNECT	用于 Web 代理服务器

在表 10-1 中，GET 是指浏览器获取由 URL 指定的服务器资源，主要用于获取由一个超文本链定义的对象。如果对象是文件，则 GET 获取的是文件内容；如果对象是程序，则 GET 获取的是该程序执行的结果；如果对象是数据库查询，则 GET 获取的是本次数据库 SQL 语句执行的结果。当客户端浏览器向服务器端传送大量的数据，并要求服务器端和公共网关接口（Common Gateway Interface，CGI）程序对数据进行进一步处理时，要使用 POST 方法。URL 是请求的服务器端资源的统一资源定位符；版本是目前 HTTP 协议通信的版本。HTTP 请求报文如图 10-12 所示。

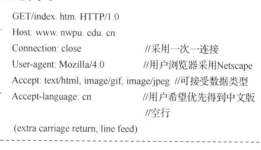

图 10-12 HTTP 请求报文

HTTP 响应报文包括开始行、首部行和实体主体。HTTP 响应报文的开始行是状态行，状态行包括 3 项内容，即 HTTP 协议的版本、状态码，以及解释 HTTP 状态码的简单短语；首部行是由若干数据字段名和对应的值组成的；实体主体一般包含一些数据，如图 10-13 所示。

在图 10-13 中，状态码都是 3 位数字，不同起始数据的状态码具有不同含义；短语是对状态码的简单解释。状态码的具体含义如下。

（1）1xx：表示通知信息，如请求收到了或正在进行处理。

（2）2xx：表示成功，如接受或执行成功。

（3）3xx：表示重定向，表示要完成请求还必须采取进一步的行动。

（4）4xx：表示客户的差错，如请求中有错误的语法或不能完成。
（5）5xx：表示服务器的差错，如服务器失效无法完成请求。

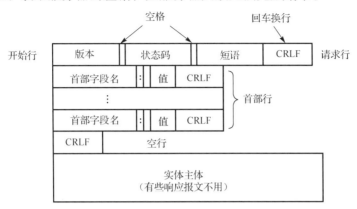

图 10-13　HTTP 响应报文格式

HTTP 响应报文如图 10-14 所示。

```
HTTP/1.0 200 OK
 ┌ Date: Thu, 06 Aug 1998 12:00:15 GMT    //发送时间
 │ Server: Apache/1.3.0 (Unix)            //服务器类型
 │ Last - Modified: Mon, 22 Jun 1998 …    //文档最后修改时间
 │ Content - Length: 6821                 //文档长度
 └ Content - Type: text/html              //文档类型
                                          //空行
   data data data data data …             //文档内容（数据）
```

图 10-14　HTTP 应答报文

如果一个用户采用 HTTP 协议访问某网站首页，则在此过程中浏览器首先分析超链接指向页面的 URL，获得服务器域名；然后浏览器向 Web 服务器发送请求，获得服务器域名对应的 IP 地址；浏览器与服务器建立 TCP 连接；浏览器发出取文件命令 GET /index.htm；最后，服务器给出响应，把文件 index.htm 发送给浏览器，TCP 连接释放。浏览器显示网站首页文件 index.htm 中的所有信息，如图 10-15 所示。

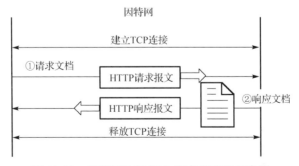

图 10-15　浏览器访问 Web 服务器流程示意图

在 Web 服务系统中，浏览器作为客户端，主要实现 HTTP 协议通信信息客户端功能；一个较为完整的浏览器结构主要由控制程序、缓存、HTTP 客户程序、其他可选服务客户

程序（如 FTP 客户端、可 HTML 解释程序、可选解释程序等）构成，如图 10-16 所示。当用户通过键盘输入有关 Web 服务请求时，如通过浏览器地址栏输入需要访问的 URL 或用户点击了某个超链时，浏览器利用控制程序，在本地缓存中查询有没有 HTTP 请求需要的内容，如果没有，则控制程序将该 HTTP 请求交给 HTTP 客户程序处理，然后 HTTP 客户程序将该请求通过网络发送给服务器。当 HTTP 客户程序接收到服务器反馈回来的 HTTP 应答后，首先在本次缓存中进行缓存，然后通过控制程序交付给 HTML 解释程序对网页标签进行解析，并将解析结果交付给显卡驱动程序显示在屏幕上。浏览器结构如图 10-16 所示。

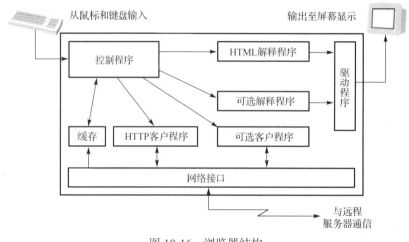

图 10-16　浏览器结构

10.2.5　Cookie 工作原理

HTTP 1.0 协议是一种无状态的协议，主要原因是服务器无法记录客户端访问服务器端资源的行为，即无法得知谁在什么时间访问了哪些网页，又进行了哪些操作。目前，一些网络应用需要对用户访问服务器端资源的行为进行跟踪，如在某网站购买图书时，服务器端需要记录用户的身份，并将该用户选择的图书放入一个"购物车"，然后集中付款，而且可以向用户推荐他可能感兴趣的图书。为了解决万维网服务的无状态问题，万维网站点可采用 Cookie 技术来跟踪用户行为，将状态信息保存在万维网服务器端，并在服务器端和客户端之间传递用户行为状态信息。

现在许多网站都支持 Cookie。例如，当用户张三第一次浏览某个支持 Cookie 的网站时，通过 3 次握手建立 TCP 连接，客户端向服务器端发送用户名及密码进行身份认证，如果身份认证通过，则服务器端会为用户张三产生一个唯一识别码（123:Cookie）作为身份识别码。服务器端采用身份识别码的目的是保护用户隐私，服务器端通过身份认证应答将该 Cookie = 123 通告给客户端。带 Cookie ID 的 HTTP 响应报文如下。

```
HTTP/1.0 200 OK
Date: Thu, 06 Aug 2013 12:00:15 GMT       //发送时间
Server: Apache/1.3.0 (Unix)               //服务器类型
Last-Modified: Mon, 22 Jun 2012 ...       //文档最后修改时间
Content-Length: 6821                      //文档长度
```

```
Content-Type: text/html                    //文档类型
Set-cookie:123                             //身份识别码
                                           //空行
data data data data data ...               //文档内容（数据）
```

张三客户端收到该响应报文，浏览器在特定 Cookie 文件（c:\document and settings\zhangsan\Cookie 文件）中增加一行：服务器域名（IP 地址）和 Cookie ID（www.amazon.cn，123）。如果张三后续继续访问该网站，则每发送一个 HTTP 请求报文，浏览器便从该特定 Cookie 文件中读取张三的身份识别码，并增加在 Cookie 首部行。带 Cookie ID 的 HTTP 请求报文如下。

```
GET /index.htm HTTP/1.0
Host:http://www.amazon.cn
Connection:close                           //采用一次一连接
User-agent: Mozilla/4.0                    //用户浏览器采用 Netscape
Cookie: 123                                //身份识别码
Accept: text/html, image/gif,image/jpeg    //可接受数据类型
Accept-language:cn                         //用户希望优先得到中文版
                                           //空行
(extra carriage return, line feed)
```

服务器利用此身份识别码，在其后台数据库中为用户张三创建一个表，记录拥有此身份识别码的用户访问网站的行为，达到跟踪该用户在网站的活动行为的目的。服务器反馈给张三的所有 HTTP 响应报文中增加的首部行中会增加一个 Set-cookie:123 字段。服务器返回的带 Cookie ID 的 HTTP 请求报文如下。

```
HTTP/1.0 200 OK
Date: Thu, 06 Aug 2013 12:00:15 GMT        //发送时间
Server: Apache/1.3.0 (Unix)                //服务器类型
Last-Modified: Mon, 22 Jun 2012 ...        //文档最后修改时间
Content-Length: 6821                       //文档长度
Content-Type: text/html                    //文档类型
Set-cookie:123                             //身份识别码
                                           //空行
data data data data data ...               //文档内容（数据）
```

服务器利用身份识别码，在其后台数据库中为用户张三创建一个表，服务器利用身份识别码，以及记录到表中的信息，可跟踪张三的访问行为，如张三访问了哪些网页、访问的顺序、进行了哪些操作、选择了哪些图书等。注意，服务器并不知道该 123 代表的是张三，只知道身份识别码 123，这是为了保护个人隐私。如果张三需要访问一个购物网站，服务器会利用身份识别码在记录表中为张三的身份识别码维护一个购物清单，这样就可以实现购物车的功能，方便张三集中付账。假如张三几天后使用同一计算机再次访问同一购物网站，由于张三的 HTTP 请求有 "Cookie:123" 字段，服务器可利用身份识别码免去张三每次登录系统的身份认证过程，并利用历史访问记录，为张三推荐其感兴趣的产品。如果张三以前在该网站使用过信用卡进行付款，则张三的注册信息包括姓名、电话、E-mail

地址、信用卡号码等，实际上这些信息已经被服务器记录并和张三的身份识别码绑定。如果张三利用同一计算机访问该购物网站，由于 HTTP 请求中具有身份识别码"Cookie:123"字段，因此服务器利用身份识别码即可验证该用户是合法用户张三，张三在登录系统或付费时不需要再用键盘输入用户名、信用卡号码，有时甚至密码也不用输入，服务器从后台数据库中利用身份识别码和注册信息已经替用户提前填写上了，用户在客户端只需点击"确认"按钮即可。但是 Cookie 技术的广泛应用也引起了争议，如系统 Cookie ID 代替了身份验证，如果谁拥有别人的 Cookie ID，那谁就可以假冒该用户登录系统，因此产生了一定的安全问题。Cookie 的应用使得用户访问网站的行为被服务器跟踪，这在一定程度上对个人隐私产生了安全风险。用户打开浏览器窗口，单击"工具→Internet 选项"选项，打开"Internet 选项"对话框，单击"隐私"选项卡，如图 10-17 所示。在该对话框中，可利用移动滑块来设置浏览器是否支持 Cookie，如果滑块移动到最上端，则表明浏览器不支持 Cookie；如果滑块移动到最下端，则表明浏览器支持 Cookie，如图 10-18 所示；如果滑块移动到中间位置，则表明浏览器部分支持 Cookie，此时用户需要设置支持 Cookie 的服务器域名或地址。

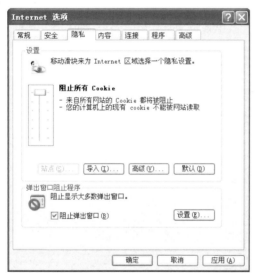 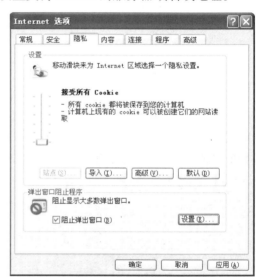

图 10-17　设置浏览器不支持 Cookie　　　　图 10-18　设置浏览器支持 Cookie

10.3　万维网文档

10.3.1　超文本标记语言

超文本标记语言（HTML）中的 Markup 就是设置标记或标签的意思。HTML 定义了许多用于排版的命令，即标签。用户在编写 HTML 文档代码时，可以使用任何文本编辑器，并将创建的文件按照 ASCII 码形式保存。HTML 文档可以用任何文本编辑器创建为 ASCII 码文件保存，仅当 HTML 文档是以 .html 或 .htm 为后缀保存时，浏览器才对此文档的各种标签进行解释，并显示在屏幕上。若 HTML 文档以 .txt 为后缀保存，则浏览器解释程序不会对标签进行解释，因此在浏览器只能看到原来的文本文件。一个简单的 HTML 文档代码如下。

```
<HTML>
<HEAD>
<TITLE>一个 HTML 的例子</TITLE>
</HEAD>
<BODY>
<H1>HTML 很容易掌握</H1>
<P>这是第一个段落，虽然很短，但它仍是一个段落</P>
<P>这是第二个段落。</P>
</BODY>
</HTML>
```

利用<IMG>标签可以在网页中插入一张图片，如图 10-19 所示。在网页中插入的图片文件为\ee\portrait.gif，图片高度为 100 像素，宽度为 65 像素。

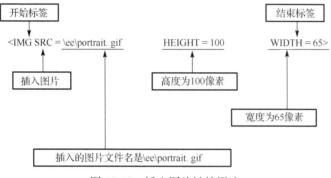

图 10-19 插入图片链接语法

万维网页面中的超链，即链接到其他网页或其他网点上的页面。超链分为本地超链和远程超链，本地超链指向本地计算机中的某个文件；远程超链的终点是其他网点上的页面。定义一个超链的标签是<A>，字符 A 表示锚（Anchor）。在 HTML 文档中，定义超链的语法示意图如图 10-20 所示。"HREF"表示超链指向的 URL，"X"表示超链在网页上显示的文本信息。

在超链示例 1（见图 10-21）中，超链的 URL 为"http://www.nwpu.edu.cn"，该超链在网页上显示的文本信息为"西工大"。在超链示例 2（见图 10-22）中，超链的 URL 为"http://www.nwpu.edu.cn"，但超链在网页上显示的是一个图片，图片名为"picture.gif"，并居中显示；如果该图片不存在，则在网页上显示文本信息"THIS IS A PICTURE"。

图 10-20 定义超链的语法示意图

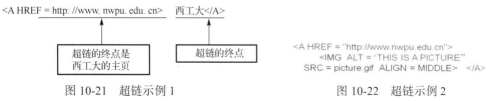

图 10-21 超链示例 1　　　　　　　　　　图 10-22 超链示例 2

10.3.2 动态文档

静态文档是指文档创建完毕后存放在万维网服务器中,在被用户浏览的过程中,文档内容不会改变。动态文档是指文档的内容是在浏览器访问万维网服务器时才由特殊应用程序,如公共网关接口(CGI)程序动态创建的。当浏览器请求到达万维网服务器时,万维网服务器要运行一个特殊应用程序 CGI,并把控制权交给此应用程序;CGI 程序对浏览器请求进行处理,并输出 HTTP 格式文档;万维网服务器把 CGI 程序的执行结果作为实体主体封装成 HTTP 响应报文,并发送给浏览器。由于对浏览器的每次请求得到的响应都是临时产生的,因此用户通过动态文档看到的内容都是不断变化的,如图 10-23 所示。

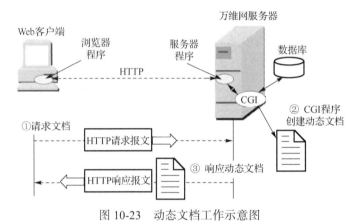

图 10-23 动态文档工作示意图

CGI 程序可以是脚本程序,也可以利用脚本语言专门编写 CGI 程序,如 Perl、JavaScript、ASP、Tcl\Tk 语言等。脚本指一个程序,它可被另一个程序解释,而不是依靠计算机的处理机来解释或执行。CGI 程序也可以用一些常用的编程语言来编写,如 C 语言、C++语言等。脚本运行起来要比一般的编译程序慢,因为它的每一条中间指令需要被解释程序处理,所以需要一些附加的指令,而不是直接由指令处理器处理。

10.3.3 活动文档

随着 Web 应用的不断发展,动态文档已不能满足实际需要,因为动态文档一旦建立,其包含的内容就固定下来无法自动刷屏变化,如无法显示一些动画效果。实际上,有两种方法可以解决浏览器屏幕无法显示连续更新的问题。

方法一:采用服务器推送,即将所有工作交给服务器负责。服务器不断地运行与动态文档关联的 CGI 程序,由 CGI 程序定期将更新信息产生的更新文档发送给浏览器。但是该方法存在以下问题。

(1)服务器中存在大量 CGI 程序同时运行,造成服务器开销大。

(2)在更新文档推送时采用 TCP 连接,该 TCP 连接在一段时间内一直保持不能释放,而且数量大,消耗了网络资源。

方法二:引入活动文档(Active Document)。活动文档把网页内容发生变化的工作交给浏览器负责。当浏览器请求一个活动文档时,服务器就会返回一段程序副本,也称为活动文档程序，并在浏览器上运行。活动文档程序可与用户直接交互，并可连续地改变屏幕显示，如图 10-24 所示。由于活动文档不需要服务器连续发送更新文档，因此对网络带宽的要求也不会太高，不会占用大量服务器资源。注意，活动文档程序本身并不包括其运行所需的全部软件，大部分的支持软件（运行环境）都由客户端浏览器自身提供。

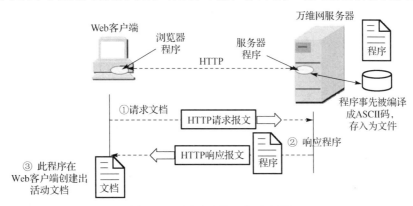

图 10-24 活动文档工作示意图

由美国 Sun 公司开发的 Java 语言可以用于创建活动文档。在 Java 语言中，使用小应用程序来编写活动文档程序。用户从万维网服务器下载嵌入了 Java 小应用程序的 HTML 文档后，可在浏览器的屏幕上点击某图像，就可以看到动画效果，在下拉菜单中单击某个项目，就可以看到计算结果。

10.4 HTTP 协议客户端实现

HTTP 协议客户端程序源代码如下。

```
#include  "stdio.h"
#include  "winsock2.h"
//WinSocket 使用的库函数
#pragma  comment(lib, "ws2_32.lib")
//定义常量
#define  HTTP_PORT   80
#define  BUFFER_SIZE   2048
#define  HOST_NAME_LEN  128

Char *http_url = "Get %s  HTTP/1.1/r/n Accept: image/jpg, image/gif,*/*\r\n"
                 "Accept-language: zh-cn\r\n Accept-Encoding: gzip,
                 deflate\r\n\"
                 "Host:%s:%d\r\n\User-Agent: browser <0.1>\r\n"
                 "Connection:Keep-Alive\r\n\r\n";

Void http_parse_request_url(const char *buf, char *host, unsigned short
                 *port, char *file_name)
{
    Int length =0;
    Char port_buf[8];
```

```
    Char *buf_end = (char *)(buf + strlen(buf));
    Char *begin, *host_end, *colon, *file;

    Begin = strchr(buf, "//");
    Begin = (begin ?  begin +2 :  buf);

    Colon = strchr(begin, ":");
    Host_end = strchar(begin, "/");

    If (host_end == NULL)
    {
        Host_end = buf_end;
    }
    Else
    {
        file = strchr(host_end,"/");
        if (file && (file +1)  != buf_end)
            strcpy(file_name, file +1);
    }
    If (colon)
    {
        Colon ++;
        Length = host_end -colon;
        Memcpy(port_buf, colon,length);
        Port_buf[length] = 0;
        *port = atoi(port_buf);
        Host_end = colon-1
    }

    Length = host_end - begin;
    Memcpy(host,begin, length);
    Host[length] = 0;
}


Int main (int argc, char **argv)
{
    WSADATA wsa_data;
    SOCKET  client_socket =0;
    Struct sockaddr_in  serv_addr;
    Struct hostent *host_ent;

    Int result =0,send_len =0;
    Char data_buf[HTTP_BUF_SIZE];
    Char host[HTTP_HOST_LEN] = "127.0.0.1";
    Unsigned short port = HTTP_DEF_PORT;
    Unsigned long addr=0;
    Char file_name[HTTP_HOST_LEN] = "default.html";
    FILE *file_web;

    If (argc != 2)
    {
        Printf ("web input:%s http://www.china.com[:8080]/default.html",argv[0]);
```

```
        Return -1;
    }

    http_parse_request_url(argv[1],host, &port,file_name);
    WSAStartup(MAKEWORD(2,0), &ws_data);


    Addr = inet_addr(host);
    If (addr == INADDR_NONE)
    {
        Host_ent = gethostbyname(host);
        Memcpy(&addr, host_ent->h_addr_list[0],host_ent->h_length);
    }
    Serv_addr.sin_family = AF_INET;
    Serv_addr.sin_port   = htons(port);
    Serv_addr.sin_addr.s_addr = addr;

    http_sock = socket(AF_INET, SOCK_STREAM, 0)
    result = connect(http_sock, (struct sockaddr *) &serv_addr,
         sizeof(serv_addr));

    send_len = sprint(data_buf, http_req_hdr_templ, argv[1],host, port);
    result = send(http_sock, data_buf, send_len, 0);

    do {
         result = recv(http_sock, data_buf, HTTP_BUF_SIZE, 0);
         if(result >0)
      {
    Data_buf[result] =0;
    Printf("%s",data_buf);
}while (result>0);

Closesocket(http_sock);
WSACleanup();

Return 0;

}

}
```

10.5 FTP 协议工作原理

10.5.1 FTP 协议概述

FTP 协议是一种基于客户/服务器模式的通信协议，它由客户端软件、服务器端软件和通信协议等 3 部分组成。FTP 客户端软件运行在用户计算机上，用户装入 FTP 客户端软件后，便可以使用 FTP 内部命令与远程 FTP 服务器端建立连接，并上传或下载文件。FTP 服务器端软件运行在远程主机上，可以为一些可信的用户注册用户账号，也可以设置一个名叫 anonymous 的公共用户账号，并对公众开放。FTP 协议在客户端与服务器端之间建立两条 TCP 连接：一条是控制连接，TCP 端口为 21，主要用于传输命令和参数；另一条是数据连接，TCP 端口为 20，主要用于传输数据。FTP 服务器端不断在一个周知的 21 号 TCP 端口上侦听用户的连接请求，当用户发出连接请求后，其控制连接便会建立起来，如图 10-25 所示；然后 FTP 客户端使用已注册的用户账号和口令，或者使用"anonymous"用户账号和"用户 E-mail 地址"口令进行登录。这时用户账号和口令通过控制连接发送给 FTP 服务器端。FTP 服务器端收到这个请求后，首先认证用户身份，然后向 FTP 客户端回送确认或拒绝的应答信息。当用户看到登录成功的信息后，便可以发出文件传输的命令。FTP 服务器端从控制连接上收到文件名和传输命令后，便在 20 号 TCP 端口上发起数据连接，并在这个连接上进行数据传输，将文件名指明的文件传输给 FTP 客户端。只要用户不使用 close 或 bye/quit 命令关闭连接，就可以继续传输其他文件。

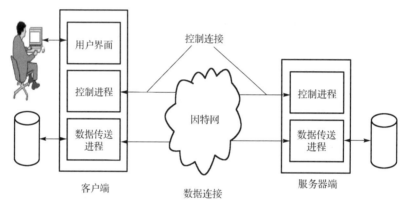

图 10-25　FTP 协议工作原理

10.5.2　FTP 协议工作模式

FTP 协议在工作时可采用两种模式：主动工作模式和被动工作模式。FTP 协议主动工作模式流程如图 10-26 所示。在图 10-26 中，FTP 服务器端首先在 21 号 TCP 端口建立控制进程，当 FTP 客户端有文件传输需求时，建立一个临时端口号为 1442 的控制进程，由客户端控制进程向服务器端控制进程发起建立 TCP 连接的请求报文，通过 3 次握手建立 TCP 连接；在已经建立的 TCP 连接上传输 FTP 协议命令，如果需要数据传输，FTP 客户端通过 PORT 命令将客户端数据进程的 1443 端口告知服务器端控制进程；服务器端数据进程在 20 号周知端口上向客户端数据进程 1443 端口主动发送连接请求，并通过 3 次握手建立 TCP 数据连接；最后客户端和服务器端在数据连接上完成数据传输。在通信结束后，通过 4 次挥手释放数据连接，但控制连接依旧保持，可以在控制连接上发送别的命令完成其他操作，直到 FTP 客户端认为本次 FTP 服务已结束，才会释放控制连接。

FTP 协议被动工作模式流程如图 10-27 所示。在图 10-27 中，FTP 服务器端首先在 21 号端口建立控制进程，当 FTP 客户端有文件传输需求时，建立一个临时端口号为 1442 的控制进程，由客户端控制进程向服务器端控制进程发起建立 TCP 连接的请求报文，通过 3 次握手建立 TCP 连接。在已经建立的 TCP 连接上传输 FTP 协议命令 PASV，表明数据传输采用被动工作模式；如果需要数据传输时，FTP 服务器端首先在 2394 临时端口上建立服务器端数据进程，并通过控制连接将服务器数据进程临时端口 2394 告知客户端控制进程。客户端数据进程在 1443 临时端口上向服务器端数据进程 2394 临时端口主动发送连接请求，并通过 3 次握手建立 TCP 数据连接，最后客户端和服务器端在数据连接上完成数据传输。在通信结束后，通过 4 次挥手释放数据连接，但控制连接依旧保持，可以在控制连接上发送别的命令完成其他操作，直到客户端用户认为本次 FTP 服务已结束，才会释放控制连接。

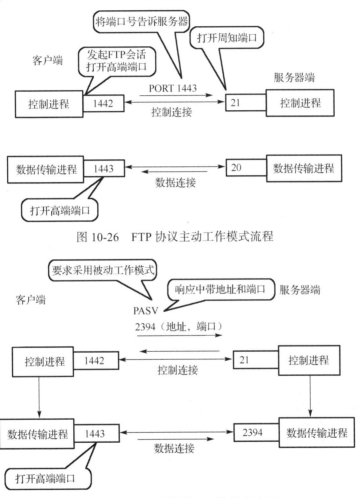

图 10-26 FTP 协议主动工作模式流程

图 10-27 FTP 协议被动工作模式流程

10.5.3 FTP 协议命令

1）打开、关闭连接

（1）FTP：建立与 FTP 服务器的会话：ftp>open Hostname [port]。

功能：在本地计算机与 FTP 远程服务器之间建立控制连接。

参数：Hostname 为服务器 IP 地址或域名，port 为控制端口号，默认为 21，可以省略。控制连接建立成功后，计算机会提示输入用户账号和密码；如果用户账号为

anonymous，则密码可用用户 E-mail 地址代替，但 anonynmous 用户只能进行文件下载，不能上传文件。

（2）close/disconnect。

功能：关闭 FTP 控制连接，回到会话状态：ftp>。

（3）bye/quit。

功能：关闭 FTP 控制连接，并结束 FTP 会话，回到 DOS 提示符状态。

2）查看信息、切换路径

（1）Pwd：显示 FTP 远程服务器当前目录。

- 举例：cd remoteDirectory。

功能：更改 FTP 远程服务器工作目录为 remoteDirectory。

- 举例：lcd [Directory]。

功能：更改 FTP 本地计算机当前工作目录为 Directory。

（2）查看信息、切换路径

- 举例：ls/dir [remoteDirectory] [LocalFile]。

功能：显示 FTP 远程服务器 remoteDirectory 目录的列表到本地计算机的 LocalFile 文件中。

remoteDirectory：如果有该参数，则将结果输出到该参数标识的文件中；如果没有该参数，则结果显示 FTP 远程服务器当前工作目录下的列表。

LocalFile：如果有该参数，则将结果输出到该参数标识的文件中；如果没有该参数，则结果显示在本地计算机屏幕上。

- 举例：mkdir Directory。

功能：在 FTP 远程服务器的当前工作目录下创建一个新目录 Directory。

- 举例：rename remotefileName newFileName。

功能：将 FTP 远程服务器上的文件的名称 fileName 改为 newFileName。

- 举例：delete/mdelete remotefile。

功能：删除 FTP 远程服务器上的 remotefile 文件。

3）对 FTP 远程服务器上的文件进行操作

- 举例：put /send/mput localfile [remotefile]。

功能：将本地文件 Localfile 上传到 FTP 远程服务器，并重新命名为 remotefile，或者保持原名称不变。

- 举例：get/recv/mget remotefile [Localfile]。

功能：将 FTP 远程服务器上的 remotefile 文件下载到本地，并重新命名为 Localfile，或者保持原名称不变。

4）其他命令

（1）! 命令。

功能：从 ftp 提示符（ftp >）临时退回 DOS 命令提示符（ >），以便运行 DOS 命令；如果要回到 ftp 提示符，则在 DOS 提示符下输入 exit 命令。

（2）?/help [command]。

功能：显示 FTP 协议所有命令说明的信息或特定命令的说明信息。

10.6　FTP 协议客户端实现

FTP 协议客户端程序源代码如下。

```
#include "conio.h"
#include "iostream.h"
#include "string.h"
#include "winsock2.h"

#pragam comment(lib, "ws2_32")

#define MAX_SIZE  4096
Char CmdBuf[MAX_SIZE];
Char  Command[MAX_SIZE];
Char ReplyMsg[MAX_SIZE];

Int nReplyCode;
Bool bConnected = false;
SOCKET SocketControl;
SOCKET SocketData

Bool RecvReply()
{
   Int nRev =0;
   Memset(ReplyMsg, 0,MAX_SIZE);
   nRecv = recv(SocketControl, ReplyMsg, MAX_SIZE, 0)
   if(nRecv == SOCKET_ERROR)
   {
      Cout<<endl << "socket recv failed!"<<endl;
      Closesocket(SocketControl);
      Return false;
   }
   If(nRecv >4)
   {
      Char *ReplyCodes = new char[3];
      Memset(replyCodes, 0,3);
      Memcpy(ReplyCodes, ReplyMsg, 3);
      nReplyCode = atoi(ReplyCodes);
   }
   Return True;
}
Bool SendCommand()
{
   Int nSend;
   nSend = send(socketControl, command, strlen(command),0);
```

```
        if(nSend == SOCKET_ERROR)
        {
            Printf("socketControl create error:%d\n",WSAGetLastError());
            Return false;
        }
        Return true;
    }
    Bool DataConnect(char * ServerIpAddr)
    {
        //向 FTP 服务器发送 PASV 命令
        Memset(command, 0, MAX_SIZE);
        Memcpy(command, "PASV", strlen("PASV"));
        Memcpy("command + strlen("PASV"),"\r\n",2);
        If(!sendCommand())
        {
            Return false;
        }
        //获得 PASV 命令的应答消息
        If(recvReply())
        {
            If(nReplyCode != 227)
            {
                Printf("PASV 命令应答错误! ");
                Closesocket(socketControl);
                Return false;
            }
        }
        //解析 PASV 命令和应答消息
        Char *part[6];
        if (strtok(replyMsg,"("))
        {
            For (int I =0; I <5; i++)
            {
                Part[i] = strtok(NULL,",");
                If(!part[i])
                {
                    Return false;
                }
            }
            Part[5] = strtok(NULL, ")");
            If(!part[5])
                Return false;
        }
        Else
```

```
    Return false

//获得FTP服务器的数据端口号
Unsigned short serverPort;
serverPort = unsigned short((atoi(part[4]) << 8) + atoi(part[5]));
socketData = socket(AF_INET,SOCK_STREAM,0);    //创建数据SOCKET
if(SocketDATA = INVALID_SOCKET)
{
    Printf("data socket creat error: %d", WSAGetLastError());
    Return false;
}

Sockaddr_in server_addr;
Memset(&server_addr, 0, sizeof(server_addr));
Server_addr.sin_family = AF_INET;
Server_addr.sin_port = htons(severPort);
Server_addr.sin_addr.s_un.s_addr = inet_addr(serverIpAddr);

//与FTP服务器之间建立数据TCP连接
Int nConnect = connect(socketData, (sockaddr *)&server_addr,sizeof(server_addr));
If(nConnect == SOCKET_ERROR)
{
    Printf("create data TCP connection error : %d\n", WSAGetLastError());
    Return false;
}
Return true;
}


Void main(int argc, char *argv[])
{
If(argc != 2)
{
Printf("please input param as the following: ftpclient  ftpIPaddr\n");
Return ;
}
If(bConnected == true)
{
    Printf("client has established the TCP control connection with server\n");
    Closesocket(sockControl)
}
Return ;
}

WSADATA WSAData;
WSAStartup(MAKEWORD(2,2), &WSAData);
```

```
socketControl = socket(AF_INET,SOCK_STREAM,0);

socketaddr_in server_addr;
memset(&server_addr,0,sizeof(server_addr));
server_addr.sin_family = AF_INET;
server_addr.sin_port = htons(21);
server_addr.sin_addr.s_un.s_addr = inet_addr[argv[1]];

int nConnect = connect(sockControl, (sockaddr *) &server_addr,sizeof(server_addr));
if(nConnect == SOCKET_ERROR)
{
    Printf("client could not establish the FTP control connection with
            server\n");
    Return;
}

//获取 Connect 应答消息
If(recvReply())
{
    If(nReplyCode == 220)    //判断应答 Code
        Printf("%s \n", replyMsg);
    Else
    {
        Printf("the reply msg is error\n");
        Closesocket(socketControl);
        Return ;
    }
}

//向服务器发送 USER 命令
Printf("FTP->USER:");
Memset(cmdBuf, 0, MAX_SIZE);
Gets(cmdBuf,MAX_SIZE)   //输入用户名并保存

Memset(command, 0, MAX_SIZE);
Memcpy(command, "USER", strlen("USER"));
Memcpy(command + strlen("USER"), cmdBuf, strlen(cmdBuf));
Memcpy(command + strlen("USER")+strlen(cmdBuf),"\r\n",2);
If(!sendCommand())
    Return;
//获得 USER 命令的应答信息
If(recvReply())
{
    If(nReplyCode == 230 || nReplyCode ==331)
        Printf("%s", ReplyMsg);
```

```
    Else
    {
        Printf("USER 命令应答错误\n");
        Closesocket(socketControl);
        Return
    }
}

If(nReplyCode == 331)
{
    //向 FTP 服务器发送 PASV 命令
    Printf("FTP > PASV:");
    Memset(cmdBuf,0,MAX_SIZE);
    For(int I = 0; i<MAX_SIZE; i++)
    {
        cmdBuf[i] = getch();  //输入用户密码
        if(cmdBuf[i] == '\r')
        {
            cmdBuf[i] = '\0';
            break;
        }
        Else
            Printf("   *   \r\n");
    }
    Memset(command, 0, MAX_SIZE);
    Memcpy(command, "PASS", strlen("PASS"));
    Memcpy(command + strlen("PASS"), cmdBuf, strlen(cmdBuf));
    Memcpy(command + strlen("PASS")+strlen(cmdBuf),"\r\n",2);

//获得 PASV 命令的应答信息
If(recvReply())
{
    If(nReplyCode == 230)
        Printf("%s", ReplyMsg);
    Else
    {
        Printf("PASV 命令应答错误\n");
        Closesocket(socketControl);
        Return
    }
}
}

//向 FTP 服务器发送 quit 命令
Printf("FTP->QUIT:");
Memset(command, 0, MAX_SIZE);
```

```
Memcpy(command, "QUIT", strlen("QUIT"));
Memcpy(command + strlen("QUIT"),"\r\n",2);
If(!sendCommand())
    Return;

//获得quit命令的应答信息
If(recvReply())
{
    If(nReplyCode ==221)
    {
        Printf("%s", ReplyMsg);
        bConnected = false;
        closesocket(socketControl);
        return;
    Else
    {
        Printf("QUIT命令应答错误\n");
        Closesocket(socketControl);
        Return
    }
}

WSACleanup();
}
```

第 11 章　IPv6 协议分析与实践

11.1　引　　言

IETF 在 1992 年提出制定下一代 IP 技术。下一代 IP 技术以 IPv6 协议为主要特征，普遍采用多协议标记交换（MPLS）技术。1992 年发表的 IPv6 协议草案（[RFC 2460]~[RFC 2463]），由于不同经济利益的竞争，目前处于草案标准阶段，主要应用策略为逐步过渡，以便有更多时间为 IPv4 协议设备升级。提出 IPv6 协议的原因之一是解决 IP 地址耗尽的问题。要解决 IP 地址耗尽的问题，可采用无类别编址 CIDR，使 IP 地址的分配更加合理；采用网络地址转换 NAT 方法以节省全球 IP 地址；采用具有更大的地址空间（128 比特位）的新版本的 IP 协议（IPv6 协议）。前两种方法只是一种过渡技术，可推后 IP 地址耗尽的时间，第 3 种方法从根本上解决了 IP 地址不足的问题。与 IPv4 协议相比，IPv6 协议仍支持无连接的传输，其主要变化如下。

（1）更大的地址空间：地址从 IPv4 协议的 32 比特位增大到了 128 比特位。
（2）采用分层地址结构。
（3）灵活的首部格式：提供了许多可选的扩展首部，可支持比 IPv4 协议更多的功能，便于协议升级扩展；提高了路由器处理效率，因为路由器不对扩展首部进行处理（除了逐跳扩展首部），仅对基本首部进行处理。
（4）支持即插即用，即 IPv6 协议地址自动配置。
（5）支持资源预分配：对多媒体通信而言，可保证一定带宽和延迟应用。
（6）安全性提高：强制使用 IPsec 安全协议，以确保数据传输的机密性和完整性。
（7）IPv6 协议首部采用 8 字节对齐，而 IPv4 协议首部采用 4 字节对齐。

11.2　IPv6 协议工作原理

11.2.1　IPv6 协议语法及语义

1）IPv6 分组结构

IPv6 分组基本首部后允许有零个或多个扩展首部，其数据部分为上层协议数据单元；所有扩展首部不属于数据部分；IPv6 分组由首部与有效载荷两部分构成，其中首部包括基本首部 + 扩展首部两部分，有效载荷 = 所有扩展首部 + 数据部分，如图 11-1 所示。

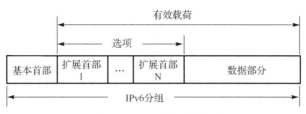

图 11-1　IPv6 分组结构图

每个 IPv6 分组都有一个基本首部，长度为 40 字节，其扩展首部可有可无，路由器只对基本首部进行处理，相对 IPv4 分组首部，其基本首部字段数减少到只有 8 个。IPv6 分组首部的特点如下。

（1）取消了首部长度字段，因为基本首部的长度是固定的。

（2）取消了服务类型字段，因为 IPv6 协议将通信量类和流标号字段结合起来，实现了服务类型字段的功能。

（3）取消了总长度字段，改用了有效载荷长度字段。

（4）取消了标识、标志和片偏移字段，该功能由扩展首部代替实现。

（5）TTL 字段改用了跳数限制字段，但作用相同。

（6）取消了协议类型字段，采用下一个首部字段。

（7）取消了首部校验和字段，加快了路由器处理 IP 分组的速度。数据链路层存在校验和，将出错的数据帧丢弃，并进行差错控制；传输层 UDP 协议有校验和，如果 UDP 用户数据报有错误，就直接丢弃，但不进行纠错；传输层 TCP 协议有校验和，可进行差错控制。由此可见，网络层差错检测是多余的，可以精简掉。

（8）取消了选项字段，用扩展首部实现选项功能。

IPv6 分组首部结构如图 11-2 所示，各字段含义如下。

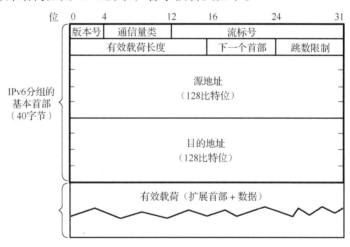

图 11-2　IPv6 分组首部结构图

（1）版本号：4 比特位，指明了协议的版本号，对于 IPv6 协议，该字段的值总是 6。

（2）通信量类：8 比特位，为了区分不同的 IPv6 分组的类别或优先级。目前正在进行不同的通信量类性能的实验。

（3）流标号：20 比特位，"流"是互联网上从特定源节点到特定目的节点的一系列分组（实时视/音频），"流"经过的路径上的路由器都可以保证指明的服务质量；所有属于同一个流的分组都具有同样的流标号，这对多媒体分组比较有用。

（4）有效载荷长度：16 比特位，指明 IPv6 分组除基本首部之外的字节数（所有扩展首部+数据部分），其最大值是 64 千字节。

（5）下一个首部：8 比特位，相当于 IPv4 协议的协议类型字段，如果无扩展首部，则与 IPv4 协议的协议类型字段一样，表明上层协议类型（TCP 协议的协议号为 6，UDP 协议的协议号为 17）；如果有扩展首部，表示后面紧跟的第一个扩展首部类型。

（6）跳数限制：8 比特位，防止分组在网络中无限期转发，类似 IPv4 协议中的 TTL；源节点在 IPv6 分组发出时设定跳数限制（255）；路由器在转发分组时将跳数限制字段中的值减 1。当跳数限制的值为零时，就要将此分组丢弃。

（7）源地址：128 比特位，分组源节点的 IPv6 地址。

（8）目的地址：128 比特位，分组目的节点的 IPv6 地址。

2）IPv6 分组扩展首部

IPv6 分组把原来 IPv4 分组首部中的选项及分片等功能都放在扩展首部中来实现，并将扩展首部留给路径两端的发送端和接收端的主机来处理。分组途中经过的路由器不处理这些扩展首部（只有逐跳选项扩展首部例外），目的是提高路由器对首部的处理效率。每一个扩展首部由若干字段构成，长度各不相同，扩展首部第一个字段为下一个首部（8 比特位），用于指明下一个首部类别，如图 11-3 所示。[RFC 2460]标准草案定义了 IPv6 分组的 6 种扩展首部，具体如下。

（1）路由选择：源路由选项。

（2）分片：大的 IPv6 分片。

（3）鉴别：身份认证。

（4）封装安全有效载荷-安全性（IPsec 协议中 ESP）。

（5）逐跳选项：为通向目的路径上的每个路由器指定分组转发参数。

（6）目的站选项。

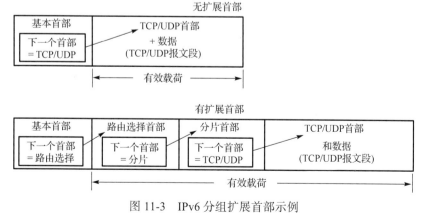

图 11-3　IPv6 分组扩展首部示例

3）IPv6 分组的分片与组装

IPv6 协议将分片与组装功能交由源节点来完成。源节点可以采用物理网络的 MTU（1280 字节），或者在发送数据前完成路径上每条链路最大传输单元，然后取最小值作为沿着该路径到目的节点的 MTU。原则上，路径中的路由器不允许进行分片；IPv6 基本首部不包含分片信息，当需要分片时，源节点在每一个分组的基本首部后插入一个分片扩展首部（8 字节）。IPv6 分片扩展首部结构如图 11-4 所示。

图 11-4　IPv6 分片扩展首部结构

（1）下一个首部：指明该扩展首部的下一个首部。
（2）保留（10 比特位）：以后使用。
（3）片偏移（13 比特位）：以 8 个字节为计算单位，表示数据部分的第一个字节在原始报文段中的位置。
（4）M：标志字段，若 M=1，则后面还有分片；若 M=0，则后面无分片。
（5）标识（32 比特位）：由源节点产生，属于同一分组的不同分片的标识相同。

在图 11-5 中，IPv6 分组的数据长度为 3000 字节。实际以太网的 MTU 是 1500 字节，分成 3 个分片，前两个分片的数据部分长度为 1400 字节，数据部分字节序号分别为 0～1399 和 1400～2799；最后一个分片数据长度为 200 字节，数据部分字节序号为 2800～2999；3 个分片的片偏移值分别为 0、175 和 350，如图 11-5 所示。

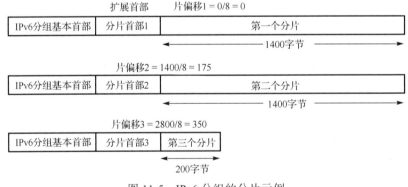

图 11-5　IPv6 分组的分片示例

11.2.2　IPv6 协议的地址空间

1）IPv6 协议的地址分类

IPv6 分组的目的地址可以是以下 3 种基本类型。

（1）单播（unicast）：传统的一对一通信方式。
（2）组播（multicast）：一对多通信方式。

（3）任播（anycast）：IPv6 协议增加的一种通信方式。任播通信的目的节点是一组计算机（所有计算机拥有相同 IPv6 地址），但分组在转发时只转发给距离自己最近的一个计算机。

2）节点与接口

支持 IPv6 协议的主机和路由器均称为节点，每一个节点可能有多个接口。IPv6 地址是分配网络给节点上面的接口的；一个接口可以分配一个单播地址，但一个节点可以有多个单播地址。任何一个单播地址都可以作为到达该节点的目的地址。

3）冒号十六进制标记法

128 比特位 IPv6 地址用 8 组 16 比特位表示，每个 16 比特位用 4 个十六进制数表示，组间用冒号隔开，具体表示方法如下。

68E6:8C64:FFFF:FFFF:0000:1180:960A:FFFF。

68E6:8C64:FFFF:FFFF:0:1180:960A:FFFF。

由于 IPv6 地址书写较长，为了简化书写，可以采用零压缩法，即一连串连续的零可以用两个冒号代替，示例如下。

FF05:0:0:0:0:0:0:00B3 可以写成 FF05::B3。

为了保证零压缩法的正确性，IPv6 协议规定：任一 IPv6 地址中只能使用一次零压缩法；数字前面连续的 0 可省略，而数字后面连续的 0 不能省略。零压缩法示例如下。

（1）1080:0:0:0:0008:0800:200C:417A 可改写为 1080::8:800:200C:417A。

（2）FF01:0:0:0:0:0:0:0101（多播地址）可改写为 FF01::101。

（3）0:0:0:0:0:0:0:1（环回地址）可改写为::1。

（4）0:0:0:0:0:0:0:0（未指明地址）可改写为::。

在零压缩法中，数字前面连续的 0 可省略，而数字后面连续的 0 不能省略；零压缩法只能使用一次。例如，点分十进制记法的后缀 0:0:0:0:0:0:128.10.2.1，使用零压缩法，可得出::128.10.2.1，此时 CIDR 的斜线表示法仍然可用。例如，IPv6 地址为 12AB:0000:0000:CD30:0000:0000:0000:0000；60 位的前缀 12AB00000000CD3 可记为 12AB:0000:0000:CD30:0000:0000:0000:0000/60，或 12AB::CD30:0:0:0:0/60，或 12AB:0:0:CD30::/60 ；如果将 12AB:0:0:CD3::/60 记为 12AB::CD30::/60，则错误。

11.3　IPv6 协议地址空间的分配

1）划分地址空间

IPv6 协议将 128 位地址空间分为两大部分：第一部分是可变长度的类型前缀，它定义了地址的网络 ID；第二部分是地址的其他部分，其长度也是可变的，可以认为由子网 ID + 主机 ID 组成，如图 11-6 所示。

2）特殊地址

未指明地址是指 16 字节全为 0 的地址，可缩写为两个冒号，记为"::"。未指明地址只能为还没有配置到一个标准的 IPv6 地址的主机当作源地址使用。环回地址是指 0:0:0:0:0:0:0:1，记为"::1"。

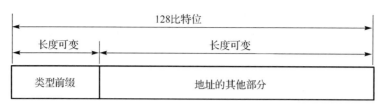

图 11-6　IPv6 地址空间结构

本地链路单播地址类似于 IPv4 协议中的私网地址，路由器不会对目的地址为私网的 IP 地址的 IP 分组进行转发。

3）IPv4 地址映射的 IPv6 地址

IPv6 地址中前缀为 0000 0000 是保留一小部分地址与 IPv4 地址兼容，因为必须要考虑到在较长的时期内，IPv4 协议和 IPv6 协议会同时存在的情况，有的节点支持 IPv6 协议，而有的节点支持 IPv4 协议。因此分组在这两类节点之间转发时，就必须进行地址的转换。图 11-7 为 IPv4 地址映射的 IPv6 地址，是一种将 IPv4 地址转换为 IPv6 地址的方法。

图 11-7　IPv4 地址映射的 IPv6 地址

5）IPv4 兼容地址

IPv4 兼容地址的前 96 位为 0，后 32 位为 IPv4 地址，用于自动隧道技术，但只有双协议栈节点才使用该地址；[RFC4291] 标准草案取消了使用该地址，在 IPv4 协议向 IPv6 协议过滤的过程中不再使用该地址。

6）全球单播地址的等级结构

因为全球单播地址的使用最多，所以 IPv6 的 1/8 地址空间为该地址。2003 年，IETF 公布了 [RFC3587] 标准草案，取消了原来划分的顶级聚合（TLA）和下一级聚合（NLA）。IPv6 地址的分级使用以下 3 个等级，如图 11-8 所示。

（1）全球路由选择前缀（001），占 48 比特位。

（2）子网标识符，占 16 比特位。

（3）接口标识符，占 64 比特位。

图 11-8　IPv6 地址的分级结构

全球单播地址的前 3 比特位为 001，因此共有 45 个比特位可分配，由互联网地址管理机构（APNIC、ARIN、LACNIC 和 RIPE）负责管理，并分配给不同组织、单位和企业，由这些组织、单位和企业决定其他部分地址的划分。单播地址主要用于公网路由选择，类似 IPv4 中的网络号。

子网标识符占 16 比特位，用于公司、单位和企业划分各自子网，对于小公司而言，也可以不用划分子网，设置 16 个 0 即可；接口标识符占 64 比特位，用于标识主机或路由器单个网络接口，类似 IPv4 中的主机号。试想，64 位 = 16 + 48（MAC 地址），这时可以不用 ARP 协议，只要知道下一跳 IPv6 地址，就知道了 MAC 地址。这样就可以将一个以太网 MAC 地址转换为 IPv6 地址。

IEEE 定义了一个标准 64 位全球唯一地址格式 EUI-64，IPv4 下 MAC 地址为 48 位，其中，前 24 位表示公司标识符，后 24 位表示公司对网络接口自主分配标识符。如果将 48 比特位转化为 64 比特位；其中前 24 位为公司标识符，第一个字节的第二低位（G\L）置为 1，表明是全球管理地址，如图 11-9 所示。

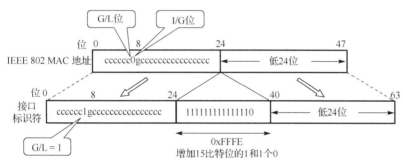

图 11-9　IPv6 地址 EUI-64 地址格式

11.4　从 IPv4 协议向 IPv6 协议过渡机制

由于因特网上存在大量的只支持 IPv4 协议的主机和路由器，因此向 IPv6 协议过渡只能采用逐步演进的办法，必须使新安装的 IPv6 系统能够向后兼容。IPv6 系统在接收和转发 IPv6 分组的同时，能够为 IPv4 分组选择路由。目前，主要采用两种过渡机制：使用双协议栈和隧道技术（[RFC2473]、[RFC2529]、[RFC2893]、[RFC3059]、[RFC4038]标准草案）。双协议栈是指在完全过渡到 IPv6 协议之前，使一部分主机（或路由器）同时装有两个协议，即一个 IPv4 协议和一个 IPv6 协议。用双协议栈实现从 IPv4 协议到 IPv6 协议的过渡，如图 11-10 所示。在此方式下，IPv6 分组首部中的某些字段可能无法恢复，这是不可避免，如流标号 X 在最后恢复出的 IPv6 分组中为空。

第二种过渡机制是采用隧道技术，如图 11-11 所示。由于源主机 A 和目的主机 F 都使用 IPv6 协议，因此源主机 A 只能向目的主机 F 发送 IPv6 数据报，路径为 A→B→C→D→E→F；因为中间路由器 B 到路由器 E 为纯 IPv4 网络，路由器 B 不能向路由器 C 直接转发 IPv6 数据报，因为路由器 C 只使用 IPv4 协议；由于路由器 B 是双协议栈路由器，可将 IPv6 首部转换为 IPv4 首部后，再将 IPv4 数据报发送给路由器 C；当 IPv4 数据报到达路由器 E（双协议栈）时，再转化为 IPv6 数据报转发给目的主机 F。

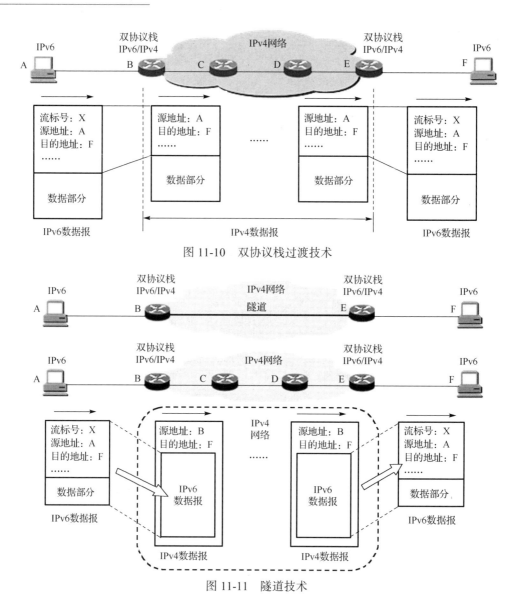

图 11-10 双协议栈过渡技术

图 11-11 隧道技术

11.5 IPv6 协议发送分组

IPv6 分组发送方实现源代码如下。

```
#include "winsock2.h"
#include "stdio.h"
#include "winsock2.h"
//加载 ws2_32.lib
#pragma comment(lib,"ws2_32.lib")
//定义 IP 分组首部数据结构
typedef struct    tagIPv6Header
{
```

```
    Union
    {
        Unsigned int version;          //版本号（4字节中前4位）
        Unsigned int priority;         //优先级（4字节中8比特位）
        Unsigned int flowlabel;        //流标号（4字节中后20比特位）
    }
    Unsigned short payloadlen;         //有效载荷长度
    Unsigned char nextheadtype;        //下一个协议类型
    Unsigned char hoplimt;             //跳步限制

    Struct
    {
        _int64 prefix;                 // IPv6 地址前缀与子网号（64位）
        Unsigned char macAddr[8];      //主机 MAC 地址（64位）
    }sourceAddr

    Struct
    {
        _int64 prefix;
        Unsigned char macAddr[8];
    }destAddr;
}IPv6Header, *ptrIPv6Header;

//64 位字节倒序，采用小端方式
_int64 hton64(_int64 host64)
{
    Char tempchar;
    Char *ptrchar = (char *)&host64;
    For(int i =0; i<4; i++)
    {
        Tempchar = ptrchar[i];
        Ptrchar[i] = ptrchar[7-i];
        Ptrchar[7-i] = tempchar;
    }
    Return host64;
}

int main()
{
    //初始化要发送的数据
    const int BUFFER_SIZE = 1024;
    char ip_buffer[BUFFER_SIZE];
    Unsigned short checkBuffer[65535];
    const char *ipv6Data = "create IPv6 packet and send!";

    //初始化 Windows Socket DLL
```

```c
WSADATA wsd;
if (WSAStartup(MAKEWORD(2, 2), &wsd) != 0)
{
    printf("WSAStartup() failed: %d ", GetLastError());
    return -1;
}
//创建 Raw Socket
SOCKET rawSocket = WSASocket(AF_INET, SOCK_RAW, IPPROTO_IP, NULL, 0,
                    WSA_FLAG_OVERLAPPED);
if (rawSocket == INVALID_SOCKET)
{
    printf("WSASocket() failed: %d ", WSAGetLastError());
    return -1;
}

//设置发送超时时间
Int timeout = 1500;
Setsockopt(RawSocket, SOL_SOCKET, SO_SNDTIMEO, (char *)&timeout, sizeof(timeout));

//设置首部控制选项，使用 IP_HDRINCL
DWORD bOption = TRUE;
int retResult= setsockopt(s, IPPROTO_IP, IP_HDRINCL, (char*) &bOption,
            sizeof(bOption));
if (retResult == SOCKET_ERROR)
{
    printf("setsockopt(IP_HDRINCL) failed: %d ", WSAGetLastError());
    closesocket(rawSocket);
    WSACleanup();
    return -1;
}
//填充 IP 分组首部各字段
IPv6Header = {0};
IPv6Header.version = 6;                 //初始化版本号
IPv6Header.version << = 8;
IPv6Header.version += 0;                //初始化 8 比特位优先级
IPv6Header.version <<= 20;
IPv6Header.version += 0;                //初始化 20 比特位流标号
IPv6Header.version = htonl(IPv6Header.version);
IPv6Header.payloadlen = sizeof(tagIPv6Header) + sizeof(ipv6Dat
IPv6Header.nextheadtype = IPPROTO_UDP;  //初始化下一个协议首部类型
IPv6Header.hoplimt = 128;               //初始化最大跳数限制

//初始化首部 128 位源 IPv6 地址
IPv6Header. sourceAddr.prefix = 0x01;   //初始化 3 位地址前缀
IPv6Header. sourceAddr.prefix <<= 45;
IPv6Header. sourceAddr.prefix += 0x01;  //填充 45 位路由前缀
```

```
IPv6Header. sourceAddr.prefix <<= 16;
IPv6Header. sourceAddr.prefix += 0x01;      //填充16位子网号
IPv6Header. sourceAddr.prefix = hton64(IPv6Header.sourceAddr.prefix);
IPv6Header.sourceAddr.mac[0] = char(0x00);   //填充8字节MAC地址
IPv6Header.sourceAddr.mac[1] = char(0x00);
IPv6Header.sourceAddr.mac[2] = char(0x80);
IPv6Header.sourceAddr.mac[3] = char(0xff);
IPv6Header.sourceAddr.mac[4] = char(0xfe);
IPv6Header.sourceAddr.mac[5] = char(0x18);
IPv6Header.sourceAddr.mac[6] = char(0x6e);
IPv6Header.sourceAddr.mac[07] = char(0xe5);

//初始化首部128位目的IPv6地址
IPv6Header.destAddr.prefix = 0x01;           //初始化3位地址缀
IPv6Header.destAddr.prefix <<= 45;
IPv6Header.destAddr.prefix += 0x02;          //填充45位路由前缀
IPv6Header.destAddr.prefix <<= 16;
IPv6Header.destAddr.prefix += 0x02;          //填充16位子网号
IPv6Header.destAddr.prefix = hton64(IPv6Header.sourceAddr.prefix);
IPv6Header.destAddr.mac[0] = char(0x00);     //填充8字节MAC地址
IPv6Header.destAddr.mac[1] = char(0x00);
IPv6Header.destAddr.mac[2] = char(0xe4);
IPv6Header.destAddr.mac[3] = char(0xff);
IPv6Header.destAddr.mac[4] = char(0xfe);
IPv6Header.destAddr.mac[5] = char(0x86);
IPv6Header.destAddr.mac[6] = char(0x3a);
IPv6Header.destAddr.mac[07] = char(0xdc);


//计算IP分组首部简单校验和字段
Memset(checkBuffer, 0, 65535);
Memcpy(checkBuffer, &IPv6Header,sizeof(IPv6Header));
Memcpu(checkBuffer + sizeif(IPv6Header), & ipv6Data,sizeof(ipv6Dara));

//组建待发送的IPv6分组
Memset(ipv6_buffer, 0, 1024);
Memcpy(ipv6_buffer, &IPv6Header, sizeof(IPv6Header));
Memcpy(ipv6_buffer + sizeof(IPv6Header),&ipv6Data,sizeof(ipv6Data));

//填充Socketaddr_in数据结构
sockaddr_in  remote_addr;
remote_addr.sin_family = AF_INET;
remote_addr.sin_port = htons(8000);
remote_addr.sin_addr.s_addr = inet_addr("192.168.0.2");
//发送IP分组
int iSendSize = size0f(IPv6Header) + sizeof(ipv6Data);
```

```
    int retResult = sendto(rawSocket, ipv6_buffer, iSendSize, 0,
        (struct sockaddr *) &remote_addr, sizeof(remote_addr));
    if (retResult == SOCKET_ERROR)
        printf("sendto() failed: %d ", WSAGetLastError());
    else
        printf("sent %d bytes ", retResult);

    //关闭 Socket, 释放资源
    closesocket(rawSocket);
    WSACleanup();
    return  true;
}
```

11.6 IPv6 协议接收分组

IPv6 分组接收方实现源代码如下。

```
#include "stdafx.h"
#include "winsock2.h"
#include "ws2tcpip.h"
#include "iostream.h"
#include "stdio.h"
#include <string.h>
using namespace std;
#pragma comment(lib,"ws2_32.lib")
//定义 IPv6 分组首部数据结构
typedef struct   tagIPv6Header
{
    Union
    {
        Unsigned int version;            //版本号（4字节中前4比特位）
        Unsigned int priority;           //优先级（4字节中8比特位）
        Unsigned int flowlabel;          //流标号（4字节中后20比特位）
    }
    Unsigned short payloadlen;           //有效载荷长度
    Unsigned char  nextheadtype;         //下一个协议类型
    Unsigned char  hoplimt;              //跳数限制

    Struct
    {
        _int64 prefix;                   //IPv6 地址前缀与子网号（64位）
        Unsigned char macAddr[8];        //主机 MAC 地址（64位）
    }sourceAddr

    Struct
    {
```

```c
            _int64 prefix;
        Unsigned char macAddr[8];
    }destAddr;
}IPv6Header, *ptrIPv6Header;


int main(int argc,char *argv[])
{
    WSADATA wsData;
    //初始化失败，程序退出
    if(WSAStartup(MAKEWORD( 2, 2 ),&wsData)!=0)
    {
        printf("WSAStartup failed\n");
        return -1;
    }

    SOCKET sock;                         //建立 Raw Socket
    if((sock=socket(AF_INET,SOCK_RAW,IPPROTO_IP))==INVALID_SOCKET )
    {
        printf("create socket failed!\n");
        return -1;
    }
    BOOL flag=TRUE;

    //设置IP分组首部操作选项，其中flag设置为true,用户可以亲自对IP分组首部进行处理
    if(setsockopt(sock,IPPROTO_IP,IP_HDRINCL,(char*)&flag,
    sizeof(flag))==SOCKET_ERROR)
    {
        printf("setsockopt failed!\n");
        return -1;
    }
    char hostName[128];
    if(gethostname(hostName,100)==SOCKET_ERROR)
    {
        printf("gethostname failed\n");
        return -1;
    }

    //获取本地IP地址
    hostent *pHostIP;
    if((pHostIP=gethostbyname(hostName))==NULL)
    {
       printf("gethostbyname failed\n");
       return -1;
    }
    //填充SOCKADDR_IN结构
    sockaddr_in addr_in;
    addr_in.sin_addr=*(in_addr *)pHostIP->h_addr_list[0];
```

```
    addr_in.sin_family=AF_INET;
    addr_in.sin_port=htons(6000);
    //把 Raw Socket 绑定到本地网卡上
    if(bind(sock,(PSOCKADDR)&addr_in,sizeof(addr_in))==SOCKET_ERROR)
    {
        printf("bind failed");
        return -1;
    }
DWORD dwValue=1;
//设置 SOCK_RAW 为 SIO_RCVALL，以便接收所有的 IP 包
#define IO_RCVALL _WSAIOW(IOC_VENDOR,1)
DWORD dwBufferLen[10];
DWORD dwBufferInLen=1;
DWORD dwBytesReturned=0;
if(WSAIoctl(sock,IO_RCVALL,&dwBufferInLen,sizeof(dwBufferInLen),&dwBufferLen,
     sizeof(dwBufferLen),&dwBytesReturned,NULL,NULL)==SOCKET_ERROR)
{
    printf("ioctlsocket failed\n");
    cout<<GetLastError()<<endl;
    return -1;
}
//设置接收数据包的缓存区长度
#define BUFFER_SIZE 65535
char buffer[BUFFER_SIZE];
//监听网卡
printf("开始解析经过本机的 IP 数据包\n");
while(true)
{
    int size=recv(sock,buffer,BUFFER_SIZE,0);
    if(size>0)
    {

        IPv6Header = (tagIPv6Header *)buffer;
        Sprintf("IP 协议版本号: %d\n", ntohl(IPv6Header.version)>>28);
        Sprintf("有效载荷长度: %d\n", ntohl(IPv6Header.payloadlen));
        Sprint("下一个协议类型号: %d\n",(int)IPv6Header.nexthead);
        Sprint("源 IPv6 地址: %H\n", IPv6Header.sourceAddr );
        Sprint("目的 IPv6 地址: %H\n", IPv6Header.destAddr );
    }
        return 0;
}
```

参 考 文 献

[1] 蔡皖东. 计算机网络[M]. 北京：清华大学出版社，2015.
[2] 谢希仁. 计算机网络（第 7 版）[M]. 北京：电子工业出版社，2017.
[3] 吴功宜. 计算机网络课程设计（第 2 版）[M]. 北京：机械工业出版社，2012.
[4] 吴功宜，董大凡，王珺等. 计算机网络高级软件编程技术（第 2 版）[M]. 北京：清华大学出版社，2011.
[5] 鲁斌，李莉. 网络程序设计与开发[M]. 北京：清华大学出版社，2010.
[6] 吴英. 计算机网络软件编程指导书[M]. 北京：清华出版社，2008.
[7] 王盛邦. 计算机网络实验教程（第 2 版）[M]. 北京：清华大学出版社，2017.
[8] 郭雅. 计算机网络实验指导书[M]. 北京：电子工业出版社，2012.
[9] 何波，崔贯勋. 计算机网络实验教程[M]. 北京：清华大学出版社，2013.
[10] 尹圣雨. TCP/IP 网络编程[M]. 北京：人民邮电出版社，2014.
[11] 杨秋黎，金智. Windows 网络编程[M]. 北京：人民邮电出版社，2015.
[12] 刘琰，王清贤，刘龙. Windows 网络编程[M]. 北京：机械工业出版社，2013.
[13] 梁伟. Visual C++网络编程案例实战[M]. 北京：清华大学出版社，2013.